Exploring

PROGRAMMABLE
ICs

Exploring

PROGRAMMABLE
ICs

Clement S. Pepper

PROMPT® PUBLICATIONS

International Standard Book Number: 0-7906-1208-9

Library of Congress Catalog Card Number: 00-110684

Acquisitions Editor: Alice J. Tripp
Editor: Will Gurdian
Assistant Editor: Kim Heusel
Typesetting: Will Gurdian
Cover Design: Christy Pierce
Graphics Conversion: Bill Skinner
Illustrations: Courtesy the author

PRINTED IN THE UNITED STATES OF AMERICA

9 8 7 6 5 4 3 2 1

Dedication

**To the creative men and women
whose ingenuity has created a need for
this kind of book.**

Contents

Preface

The Purpose of This Book

A programmable integrated circuit (IC) is a device designed for operation under program control from a computer or microprocessor. A typical device is provided with a bidirectional bus for inputting instructions and data and outputting data, control lines for read/write and device selection, and addressing lines for internal function definition.

The requirement for computer control imposes a burden for many hobbyists, experimenters, and others who may have an interest in evaluating the functions and operation of the device. This is because of the requirement for a suitable computer and the need for software tailored to the device. It would be convenient to have functions equivalent to software available in hardware circuitry capable of operating the device on the bench for observation and learning of its performance, thereby bypassing the software constraint.

In this book you will find circuits and procedures for the exploration of a variety of programmable ICs. For the most part, operation of the device requires little more than readily available, affordable breadboarding materials and circuits: simple clocks, shift registers, bounceless switches, monostables, and LED status indicators. These parts are assembled as standard reusable modules that are readily adopted for use with any of the devices described in the book. Operating procedures for bench-top exploration are included with each chapter. For most devices, a simple LED will suffice for status indication. Where fast pulse or waveform generation exists, an oscilloscope will be required.

In these chapters you will discover operating setups for a variety of programmable devices ranging from D-A converters to universal asynchronous receiver/transmitters (UARTs), interval timers, peripheral interface adapters (PIAs), and others. With these, performance is easily observed as the device is put through a sequence of operations. In addition to its value for learning, this could also be a possible means for checking the performance of a suspect device in some circumstances. The devices chosen reflect those thought by the author to be of the most interest. Experience gained from these can be applied to those of your particular needs.

The operating modules are described in the opening chapter. Beyond their educational value for some devices, this approach could lead to useful stand-alone instrumentation applications—a specific example being the 8253 or 8254 programmable interval timer, the possible basis for a utility multifunction waveform generator.

A Look Inside

Chapter 1: What We Need To Get Started. We start off with the construction of a few simple modules. The first—the "key," as it were—of these is based on a serial input, eight-bit parallel output shift register: the 74C164. Other components are toggle switches, a Quad NOR as a pulse source, an octal latch, the 74HCT373, and bounceless switch circuits with LED status indicators. Other module needs are met with a two-switch, dual monostable assembly; four-DIP and eight-DIP switch assemblies, and a couple assemblies of LEDs. All are low cost, easy to make, and simple to use, as we will see when we make use of them in the chapters that follow.

Chapter 2: The Analog Devices AD588 "DACPORT". The AD588 is an eight-bit digital-to-analog converter (DAC) providing reference, output amplification, and data latch on a single chip. It operates from a single 5V power source. It was designed with most of the available microprocessors, including the 8080A, Z80A, 1802, 6800, and others in mind. The modules we construct in Chapter 1 will allow us to simulate these with simplicity and ease.

Chapter 3: The TR1602/AY-5-1013 UART. This is a general purpose programmable MOS/LSI device for interfacing the asynchronous data output of a peripheral with the parallel data input requirement of a computer or terminal. In practice, two UARTs are employed, with or without an intervening modem. The transmitter section accepts parallel data, which is converted to serial for sending. The receiving section converts its serial input to a parallel output format. The transmitted serial string is provided with start, data, parity, and stop bits. Both transmitter and receiver are double buffered.

Word length may be five, six, seven, or eight bits. Parity may be odd or even. We will provide a clock, a source for the parallel input, and indicator LEDs, and we will also control the data transmission and observe the transition back to parallel.

Chapter 4: The SY6522/SY6522A VIA. This device, originally designed to work with the 6502 microprocessor, is appropriately named. Its primary features include two eight-bit I/O ports with individually selectable bidirectional lines, two 16-bit programmable counter/timers, a serial data port, and latched output and input registers. The device is TTL compatible with CMOS-compatible peripheral control lines. Expanded "handshake" capability provides for positive control of data transfers between processor and peripheral devices. Operation is at 1 MHz (SY6522) or 2 MHz (SY6522A).

Chapter 5: The R6520/MC6820/MC6821PIA. These devices provide the means for interfacing to the 6502 and M6800 family of microprocessors. Their main features include an eight-bit bidirectional data bus, two programmable control registers, two programmable I/O data-direction registers with individually selectable bidirectional lines, four individually controlled interrupt input lines (two of which are usable as handshake controls for input and output peripheral operation), and CMOS-compatible peripheral control lines with TTL port buffer-drive capability.

Chapter 6: The INS8250/INS8250-B, NS16450, INS8250A, NS16C450, and INS82C50A UARTs. These devices function as a serial data input/output interface in a microcomputer system. The UART performs serial-to-parallel conversion on data characters received from a modem or peripheral device. Parallel-to-serial conversion is performed on data characters from the CPU. Status information on transfer operations is reported, as are errors in parity, overrun, framing, or break interrupt. A baud rate generator is included. A 16X clock is internally generated for driving internal transmitter and receiver logic. Complete modem control is provided. Three-state TTL compatibility is provided for the bidirectional data bus and control lines.

Chapter 7: The NS16550/NS16550A/NS16550AF UART With FIFOs. This device is an improved version of the 16450 UART and is capable of running on existing 16450 software. On power-up, the two are functionally equivalent. The 16550AF, however, can be put into an alternative FIFO mode that serves to relieve the CPU of excessive overhead. This chapter illustrates the FIFO characteristics of this device, as well as its 16450 performance.

Chapter 8: The 8251/8251A USART. The 8251A is an enhanced version of the 8251, designed for data communications with Intel's microprocessor families. As a peripheral device, it's capable of using virtually any serial data transmission technique in use. Data is accepted in parallel form and converted to serial for transmission. It can also receive serial data streams simultaneously and convert them to parallel. Both synchronous and asynchronous operation are possible. Parity, overrun, and framing errors are detectable. All inputs and outputs are TTL compatible.

Chapter 9: The 8253/8253-5 Programmable Interval Timer. This device is a programmable counter/timer designed for use as an Intel microcomputer peripheral. It employs NMOS technology, operating from a single +5V supply. Its role is the generation of accurate time delays, thereby minimizing software overhead. Other capabilities include a programmable rate generator, event counter, binary rate multiplier, real-time clock, digital one-shot, and complex motor controller. Interfacing is via a three-state, bidirectional system data bus.

Chapter 10: The 8254/8254-2 Programmable Interval Timer. The 8254 is a superset of the 8253 and is compatible with most microprocessors. It features three independent 16-bit counters offering six programmable counter modes. Frequency range of the 8254 is DC to 8 MHz, to 10 MHz for the 8254-2. The device features a status read-back capability not provided with the 8253.

About the Author

Clement S. Pepper has had a very diverse and fascinating career, beginning with the U.S. Navy in 1944 where he was schooled in Naval Fire Control and served for three and one-half years on the U.S.S. Harwood, DD861 and 16 months on LST1123. He left the Navy with the rating of Fire Controlman, first class. During this time, Pepper had 16 weeks of A school, eight weeks of operational school, and 33 weeks of advanced Fire Control school. He had considerable experience with five-inch and 40-mm gunfire control systems. It was his Navy experience that made him decide to pursue an engineering career.

In 1952, Pepper became an electrical drafter at Ryan Aeronautical in San Diego, where he rose to the position of design engineer. He worked on the electrical design of the Firebee Target Drone. For two years, Pepper was on assignment to Douglas Aircraft in Santa Monica for the development of the power plant of the DC8 aircraft. His task for this was design of the electrical installation on the J57 engine.

In 1960, Pepper moved to the Scripps Institution of Oceanography's Marine Physical Laboratory and worked as an assistant engineer with only two years of engineering at San Diego State University under his belt. During the next 10 years, he developed an operational sea-floor three-component fluxgate magnetometer system for recording minute variations in the Earth's magnetic field. He also earned a bachelors degree in physics with a concentration in electronics from San Diego State University (SDSU). He continued his education through the University of California Extension with EE courses, and later 15 additional EE units at SDSU.

In 1970, Pepper left Scripps for employment at Gulf General Atomic, where he spent the next five years on nuclear weapons systems testing conducted at the Nevada Test Site. In 1973, Gulf sold his division to Intelcom Rad Tech and it became IRT. Following his first nuclear event, Pepper became the project officer on six more, culminating with the seventh in 1975. At that time, Pepper was assigned to the company's commercial products division. From 1976 through much of 1979, he was the manager of engineering and manufacturing. During his 19 years with the company, Pepper participated in the development of a variety of systems employing penetrating radiation with a variety of sources—nuclear, X-ray, and gamma ray—for various companies.

From 1978 to 1981, Pepper returned to SDSU part time as an English major, taking courses in literature and creative writing. Throughout his career, he has written numerous articles for hobby electronics and computer magazines. More recently, Pepper has authored three books, including *Build Your Own Home Lab* and *The Digital IC Gallery,* both from PROMPT® Publications.

Chapter 1
What We Need To Get Started

Introduction

Programmable ICs are designed to function under software control—which is all well and good; that's how it should be. It does, however, present a problem to those of us who might wish to explore the devices' behavior without having to write software and configure our computer for each device of interest. Manufacturers' data books, for the most part, do not appear to be written by English majors, and an alternative method for their exploration would seem to be desirable.

This chapter covers the development of circuitry that I have found useful for the control and response display for a variety of programmable ICs. The selections are of those most likely to be of interest. From these, we can "leapfrog"—as it were—to other devices of specific interest to ourselves.

First, let's think about the circuit requirements for our hardware controller. Our typical device will feature an eight-bit-wide bidirectional data bus. In addition to data, it may function in a control mode as well. Most likely there are one or more device and control selects, along with read and write enables. Addressing inputs, where required, vary depending on the device. And then there's the need for a clock. For some, such as the 8253 and 8254, a gate enable is needed. Other enable needs exist and vary between devices.

So much for inputs. What about responses? An advantage offered by our approach is that we can slow things down, as with a single-step clock for most parts. This allows us to use LED lights as status indicators for many functions. In some cases, we may have to fall back on the oscilloscope, but whenever possible it's best to circumvent that.

One point of importance is that, since the operations are conducted in slow motion, so to speak, time response cannot be observed. But the point of this book is to understand how the device goes about its operations, which we can accomplish at a reduced speed.

In working out solutions to those needs we have to deal with, let's begin with the data bus, which has the most demanding requirements. Basically, the need is to set each line—or bit, if you will—to a specific value. The Programmable Device

Under Test (PDUT, as I may refer to it) normally doesn't respond to whatever is sent out onto the bus until given a write (/W) instruction (in this book, "/" indicates active low). Where the device response is to be output on the data bus, a read (R) instruction is required. Since the two cannot occur simultaneously, we can toggle between them thusly: R//W.

We might think the most direct approach to setting the data bus lines to be eight high-low toggle or DIP switches. An LED tied to each switch would remind us of each line's status. That's fine, if you care to take that route. I didn't, as I felt an array of switches to be confusing and time consuming. Instead, I went for a serial input, parallel output, static-shift register (SSR) approach. The SSR used is the 74C164.

Details of this CMOS IC are shown in *Figure 1.1*. By clocking a sequence of eight high and/or low states at the input, the eight lines are set from a single switch. LEDs on the outputs provide visual reminders as the high-lows are "walked" through the register. Note that the flip-flops are D-type. With these, the "Q" output state is that of the "D" input. Initially, we may think it strange to set bits in this manner, but it won't take very long to see how slick this really is.

Simply dropping the switch states on the bus lines is a no-no for the simple reason that, for most of these guys, data is fed back from the PDUT. So we have to have a means of blocking this. A suitable device is the three-state 74 HCT373 latch. Details are given in *Figure 1.2*.

The 74HCT373 is a good choice, as it is able to drive both TTL and CMOS loads. The value of the three-state is the high impedance presented when reading the bus, as we will see. With the latch held open, our inputted data flows on through.

The most direct approach for reading from or writing to the data bus is to set up a bounceless two-position switch, with the high position for a read and the low for a write operation. *Figure 1.3* shows a generic bounceless switch using two two-input NAND gates. The logic 1 output is assigned to read, while the logic 0, to write. If we normally keep the switch in the read position, then we can toggle it to enable the write position. (This will work for most, though not for every, device in this book.)

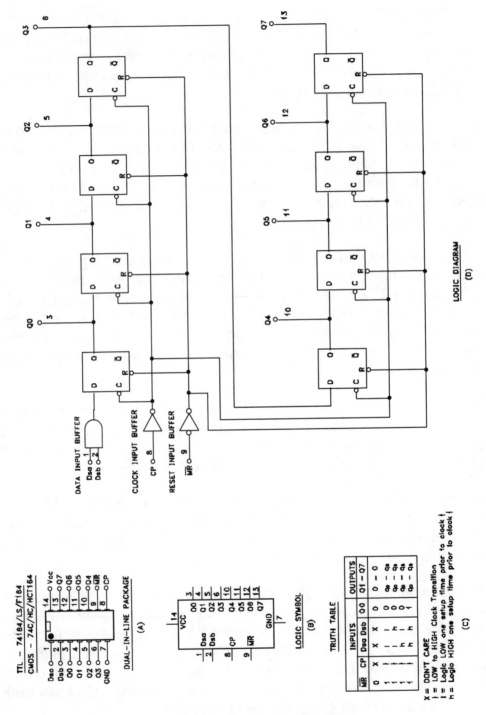

Figure 1.1. The 74C164 eight-bit serial-in, parallel-out static shift register.

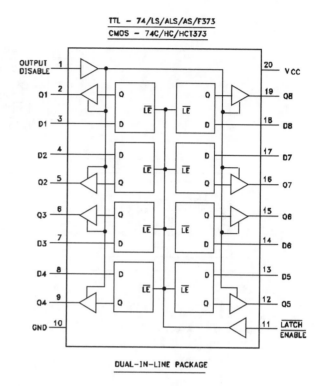

TTL — 74/LS/ALS/AS/F373
CMOS — 74C/HC/HCT373

DUAL-IN-LINE PACKAGE

Figure 1.2. The 74n373 TRI-STATE Octal D-Type latch. TTL and CMOS family devices have identical pinouts and logic functions.

OUTPUT DISABLE	LATCH ENABLE	D	Q
L	H	H	H
L	H	L	L
L	L	X	Q
H	X	X	HI-Z

TRUTH TABLE

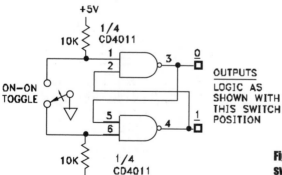

OUTPUTS
LOGIC AS
SHOWN WITH
THIS SWITCH
POSITION

Figure 1.3. The circuitry for a generic bounceless switch using two gates of a QUAD 2-input NAND gate.

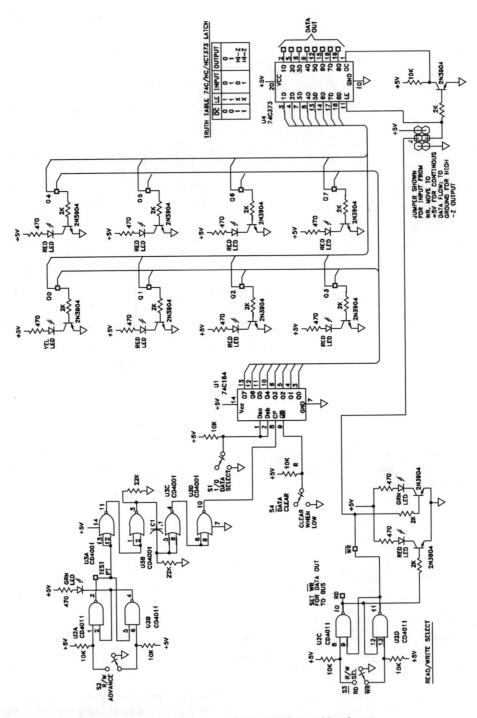

Figure 1.4. The shift register-based circuit for eight-bit data/control bus input.

Operation of the read/write switch is seen in *Figure 1.4*. The circuitry of Figure 1.4 is the engine driving the explorations we'll do in the chapters that follow.

The switch RD contact connects to the latch enable (LE) terminal of a 74HCT373, while the WR connects to its output control (OC). The circuitry is shown for S3 in the diagram. The actual data to be placed on the bus is derived from the actions of switches S1 and S2. Switch S1 selects the logic level (0 or 1) to be placed on the bus, and S2 advances the shift register. We keep track of which bit is at what level by the LED status. When all eight display the correct logic, we toggle switch S3 to put the 74HCT373 in the write mode. The operation will become clearer once we move on to a chapter application.

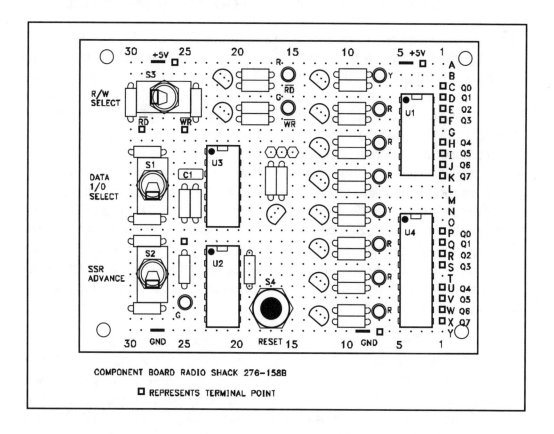

Figure 1.5. The author's construction of the shift register-based circuit for eight-bit data/control bus input.

Figure 1.5 shows the board layout I used. It's not necessary for you to duplicate this as shown, but some features will be helpful to our under-standing of the uses to which it will be put.

Let's go back momentarily to the circuit schematic. In the upper left we see switch S2, CLOCK IN, and the bounceless switch circuit for advancing the shift register logic. The LED is just a status indicator. The circuits in this book are loaded with LEDs as an aid to keeping track of the action. The four NOR gates with their resistors and capacitor form a one-shot, deliver-ing a pulse on the order of a couple of milliseconds. This is the clocking input to the 74C164.

Switch S1, DATA SET, provides a logic "1" or "0" for entry into the shift regis-ter on each toggle of switch S2. Switch S4 clears the register flip-flops.

Throughout this book, there will be examples of LED use for status indica-tion. Many will include a transistor driver for needed inversion. Sometimes the driver will be present to prevent loading of a source. Each output of the 74C164 drives the base resistor of an NPN transistor switch. With the input high, the switch is "closed" and the LED is on, indicating the register output is a logic 1. When we view the construction (see Figure 1.5), we see these LEDs positioned in a vertical column. The first (top) and fifth LEDs are yel-low as an aid to reading the value being sent out to the bus. The remaining LEDs are red. The trick to setting the bits is to follow the sequence from the top down as we toggle S2. Always begin with a logic 1. The top LED repre-sents the low-order 0 (LSB) bit at the output, the seventh (MSB) at bottom. The register outputs also connect to the inputs of U4, the 74HCT373. Two sets of output posts are shown: the upper eight connect to the register out-puts; the lower set connects to the latch outputs. This allows for greater flexibility in driving external circuits.

Now we will consider additional circuit assemblies that are needed to round out the device operations. *Figure 1.6* shows the circuitry and construction used for two bounceless switches and two monostables. They're shown together as they were constructed side by side on a single convenient Radio Shack construction board. The monostables are the 74C221: the first half sends the pulse out; the second drives a status LED. If you won-der at this, it's comforting to know the pulses are going out. A drawback to

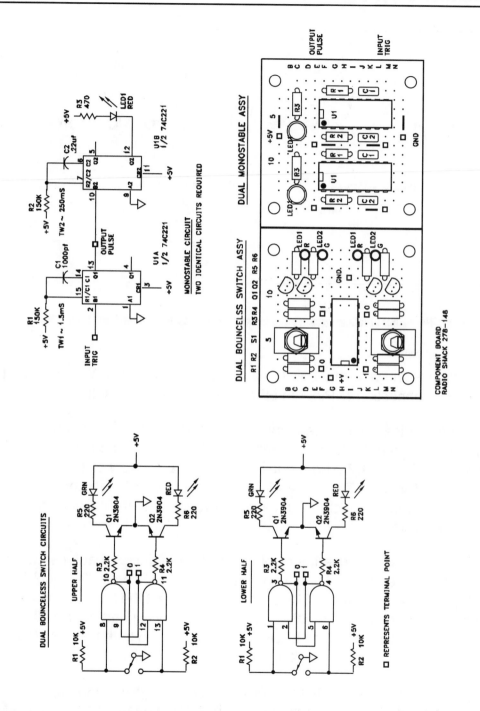

Figure 1.6. The author's circuitry and construction of a dual bounceless switch and dual monostable assembly.

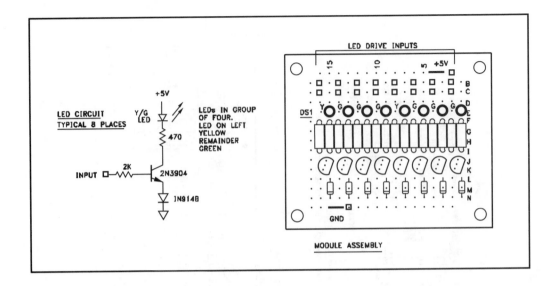

Figure 1.7. The author's module circuitry and construction for eight independent status LEDs.

the circuits shown is that only the positive pulse goes out. You may want to make a connection to Pin 4 to obtain this for your own setup.

The need arises for independent display modules. In fact, for most devices there will be a need for more than one. The circuitry and construction are shown in *Figure 1.7*. The module is seen to simply consist of eight independent displays. The first and fifth LEDs are yellow—in contrast to the remaining LEDs, which are shown as green. The choice of colors is, of course, yours.

We cannot always drive the display without a buffer, and at times we may want to isolate the display during a test. *Figure 1.8* is the display of Figure 1.7 preceded by a 74HCT373 latch. We can select the latch status with the two jumpers shown. The latch is handy when we need to read the data being output on a bus. The LEDs can be connected to the latch outputs, or used independently as desired.

Usually, we will simply let the data flow through, though the drawing shows the jumpers positioned for external control. The two jumpers must be coordinated.

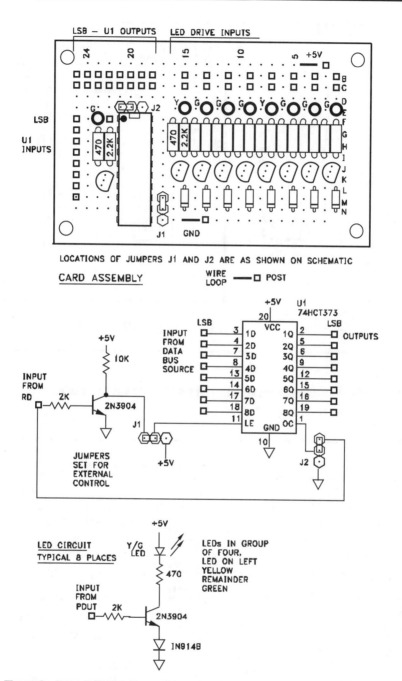

Figure 1.8. The author's module circuitry and construction for an auxiliary 74HCT373 latch plus eight independent status LEDs. This module is derived from the LED module of Figure 1.7 and differs from it only in the addition of the latch circuitry.

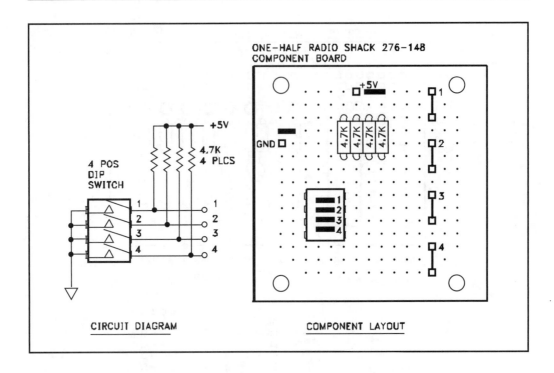

Figure 1.9. The author's circuity and construction for a 4-SPST DIP-switch module.

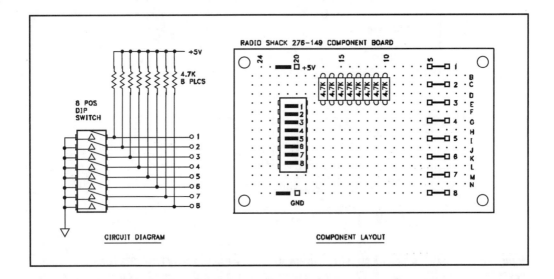

Figure 1.10. The author's circuity and construction for an 8-SPST DIP-switch module.

Two DIP-switch modules provide for miscellaneous control and addressing requirements. A four-position DIP module is described in *Figure 1.9*; an eight-position module in *Figure 1.10*.

For some situations, such as bidirectional peripheral ports, there is the need for circuitry to pass or block the flow of data between a source, typically high/low logic switches and the port. *Figure 1.11* describes the circuits and layout on a prototyping strip for this purpose.

Figure 1.12 illustrates the application of these modules in combination with solderless experimenter strips. This is an active layout employed in

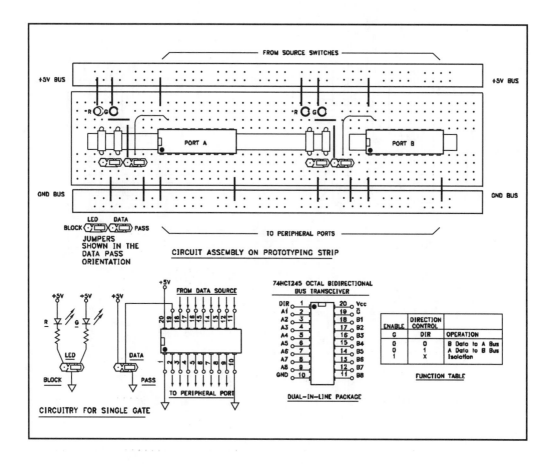

Figure 1.11. Circuitry and the module configuration used for passing or blocking eight-bit data flow between peripheral ports and source switching.

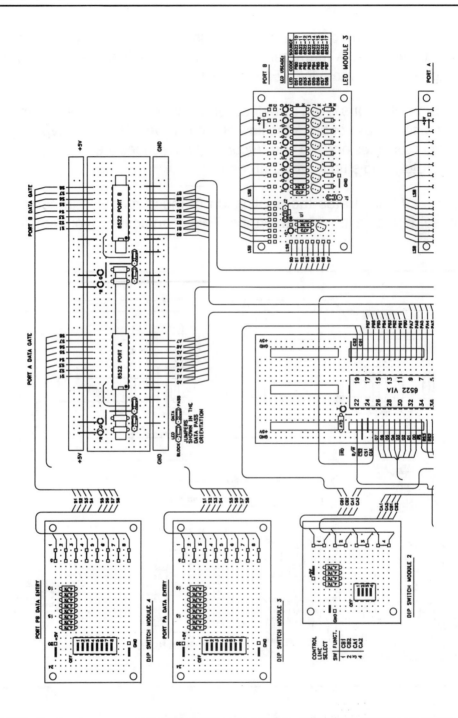

Figure 1.12. *(Continued on next page.)* **An example taken from Chapter 4 of how the modules are used in our investigations of device performance.**

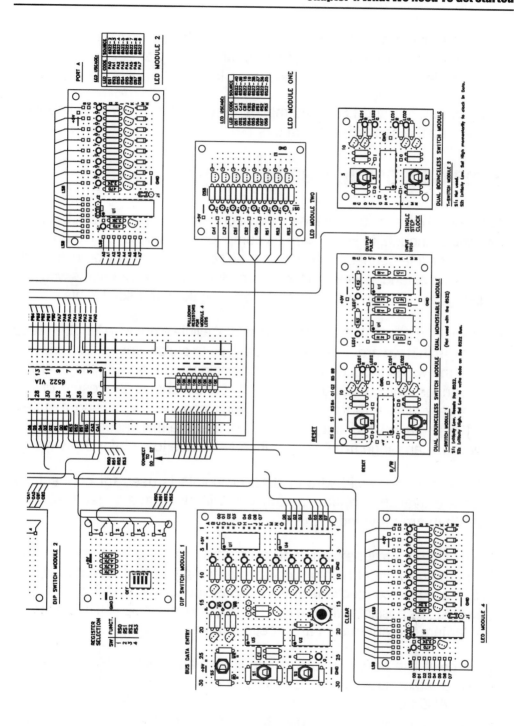

Chapter 4 with the 6522 versatile interface adapter (VIA). It may look a little frightening at this point, but it's relatively easy to set up and, with a little experience, the operation becomes familiar.

Additional switch and LED assemblies may be described in subsequent chapters as the need for them arises.

Some Construction and Use Tips

You may not care to duplicate the module constructions as shown. It is to our advantage, though, to keep them as compact as possible to hold down the length of the interconnecting wires. I haven't encountered any problems arising from the extended wiring in the layout. But I have had problems with poor soldering, so I caution that care be taken to make good solid joints.

When making connections, don't be afraid of a little extra wire length—keeping in mind that we are operating the devices at a slow speed. You will want to route the wiring around the LEDs and avoid interfering with the operation of the various switches. Double check all connections, as the results of such can be very difficult to track down at worst and yield confusing results at best. Keep in mind, too, that the effects of a single misplaced wire or faulty connection can range from a minor inconvenience to a major disaster.

AWG 22 insulated solid wire of differing colors is used for connections. These are available in kits, though you will also have a need for wires longer than are provided in the kits.

A super strip or similar experimenter socket is used as a centerpiece for the PDUT. You may find it helpful to plug in an empty socket as a placeholder while making the connections and initial power on tests. Check out as much of the setup as possible with the PDUT absent to minimize risk to the device. A 6V lantern battery makes an excellent power source.

At this point, the best way to continue is with an example. For this, I have chosen a rather straightforward device to begin our work with: the Analog Devices AD588, a programmable digital-to-analog converter (DAC) designed for use with a variety of microprocessors. We'll check it out in the next chapter.

Chapter 2
The Analog Devices
AD588 "DACPORT"

Introduction

The AD588[1] is an excellent device for us to begin our exploration, as it is relatively easy to set up and follow its operation. This is not to imply that the AD588 is short on complexity, which is hardly the case. This is a microprocessor-compatible, eight-bit digital-to-analog monolithic converter (DAC). A popular application of a DAC is the conversion of an eight-bit word placed on the bus to an analog voltage.

The AD588 is a complete eight-bit unit, provided with an internal 1.2V bandgap reference, an output amplifier, and data latch. It's designed to operate with a single positive supply voltage over the range of +4.5V to +16.5V. Here, we use a 5V supply, which enables a full-scale output of 2.56V at a resolution of 10 mV per bit. The chip is provided with laser-trimmed, thin-film resistors for absolute accuracy and linearity. The circuitry is internally compensated for minimum settling time on both ranges: typically to ±½LSB for a full-scale 2.55 step in 800 nS. Digital input currents of 100 µA max minimize bus loading.

The eight-bit input register and fully microprocessor-compatible control logic allow direct connection to eight-bit or 16-bit data buses with operation from standard control logic. The latch may be disabled for direct DAC interfacing.

Figure 2.1 provides an overall description of the DAC, with package details, function diagram, and a partial listing of input/output values. The figure also details power connections and two configurations: for the range of 0 to 2.56V and a range of 0 to 10.0V.

Figure 2.2 shows the circuit connections for our bench-top operation. The schematic for the shift-register-based control (see *Figure 1.4*) is included to provide a complete overview of what is transpiring. Note that the outputs of the 74HCT373, U4, are also connected at the DAC inputs to the LED inputs of the latch and LED card. The indicators on this card should match those on the shift-register card—just a touch for added assurance that the DAC is receiving the expected digital input.

Figure 2.3 shows my bench setup (see Figures 1.5 and 1.8 for the card assemblies). Here, the DAC is mounted on a breadboard with connecting

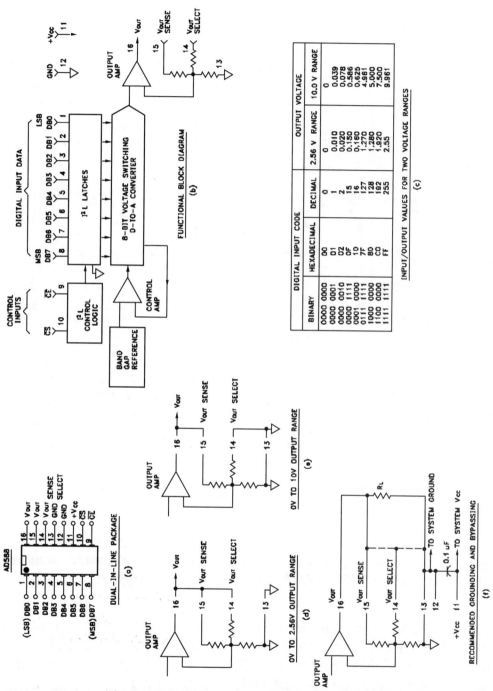

DIGITAL INPUT CODE			OUTPUT VOLTAGE	
BINARY	HEXADECIMAL	DECIMAL	2.56 V RANGE	10.0 V RANGE
0000 0000	00	0	0	0
0000 0001	01	1	0.010	0.039
0000 0010	02	2	0.020	0.078
0000 1111	0F	15	0.150	0.586
0001 0000	10	16	0.160	0.625
0111 1111	7F	127	1.270	4.961
1000 0000	80	128	1.280	5.000
1100 0000	C0	192	1.920	7.500
1111 1111	FF	255	2.55	9.961

INPUT/OUTPUT VALUES FOR TWO VOLTAGE RANGES

(c)

Figure 2.1. DIP package and function details of the AD588 DAC. The circuit connections for two output voltage ranges are given. The table relates several output voltages to digital inputs.

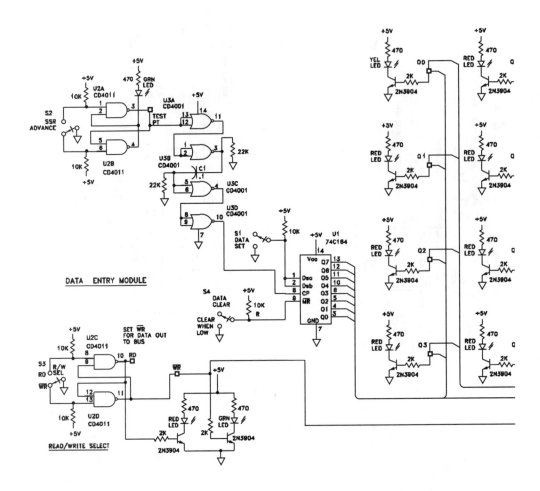

Figure 2.2. *(Continued on next page.)* **The circuit connections for the bench-top operation. The shift register circuitry is included to provide insight into the data-bus control functioning.**

jumpers, the filter capacitor, and power connections. A multimeter, preferably digital (not shown), connects to pin 16. As we vary the input bits, we can verify the analog voltage tracking. *Table 2.1* illustrates several expected output values based on the shift-register LED settings.

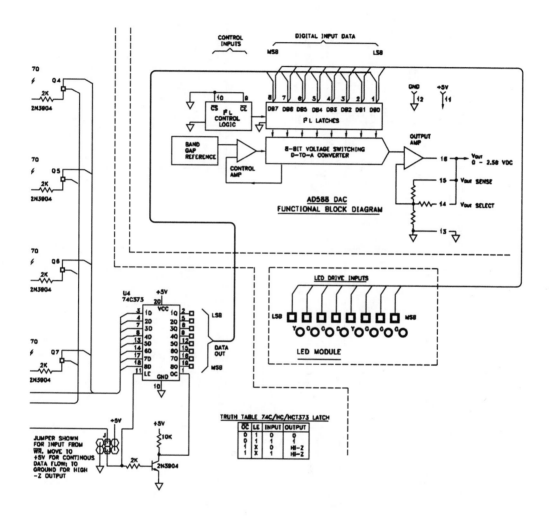

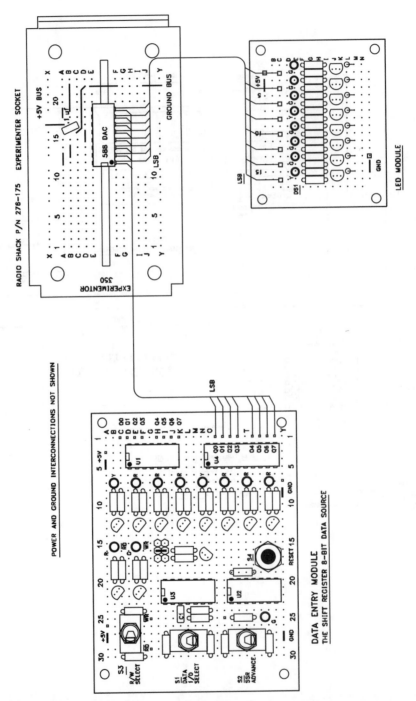

Figure 2.3. The circuit card layout for the bench-top operation. The module assemblies make for a quick and efficient interconnection.

LED	BINARY VALUE	INC SUM (mV)	LED	BINARY VALUE	INC SUM (mV)	LED	BINARY VALUE	INC SUM (mV)	LED	BINARY VALUE	INC SUM (mV)	LED	BINARY VALUE	INC SUM (mV)	LED	BINARY VALUE	INC SUM (mV)
Y●	001	0010	Y○			Y●	001	0010	Y○			Y○	001		Y○	001	
R●	002	0030	R●	002	0020	R○			R○			R○	002		R○	002	
R●	004	0070	R○			R●	004	0050	R●	004	0040	R○	004		R●	004	0040
R●	008	0150	R○			R●	008	0130	R●	008	0120	R●	008	0080	R○	008	
Y●	016	0310	Y●	016	0180	Y○			Y○			Y●	016	0240	Y○	016	
R●	032	0630	R●	032	0400	R●	032	0450	R●	032	0440	R○	032		R●	032	0360
R●	064	1270	R○			R●	064	1090	R○			R●	064	0880	R○	064	
R●	128	2550	R●	128	1780	R○			R●	128	1720	R●	128	2160	R●	128	1640

○ INDICATES BINARY 0
● INDICATES BINARY 1

Table 2.1. A sampling of analog output voltages for the given digital inputs. The LED colors relate to the "LED module" (ref. Figure 2.2).

References

1. Analog Devices, "DACPORT Low Cost, Complete µP-Compatible 8-Bit DAC, *Data Converter Reference Manual,* Volume 1, 1992, pp. 2-47.

2. Reference for Figure 2.1: Ibid. Figures 2, 3, and 5.2, and Table: "Input Logic Coding."

3. Reference for Figure 2.2: Ibid. Figure 2.

Chapter 3
The TR1602/AY-5-1013 UART

Introduction

An asynchronous receiver/transmitter (UART) is a programmable MOS/LSI device for providing serial data transmission between a computer and its peripherals. The device accepts a parallel input from the source, converts it to an asynchronous serial format, and transmits it to a second device where the inverse transformation to the parallel format takes place. Inputs and outputs are TTL compatible.

The TR1602/AY-5-1013[1,2] has applications for peripherals, terminals, minicomputers, modem operations, data multiplexers, card and tape readers, printers, encoders, and remote data-acquisition systems. It is a relatively easy device to program.

Internally, the device is organized into two independent sections: the transmitter and the receiver. *Figure 3.1* provides DIP packaging pinouts and an overall function diagram. *Figure 3.2* shows a detailed block diagram of the transmitter section with a timing diagram. *Figure 3.3* shows a detailed block diagram of the receiver section with a timing diagram.

The transmitter section converts the parallel data input to a serial word format. Along with the data, this section also provides start, parity, and stop bits as specified. Word length may be five, six, seven, or eight bits. Parity may be odd or even. There may be one or two stop bits, with 1.5 when transmitting a five-bit word length.

The receiver section restores the received serial word to parallel form. In this process, it checks for correct start and stop bits, and errors in parity, framing, or overrun conditions. The receiver and status flag outputs are three-state.

Table 3.1 defines the device inputs and outputs for each of the 40 pins. Selection of the various operating modes is by the coding of five control inputs. These are defined in *Table 3.2*. Relate the definitions of Table 3.2 with the definitions of Table 3.1.

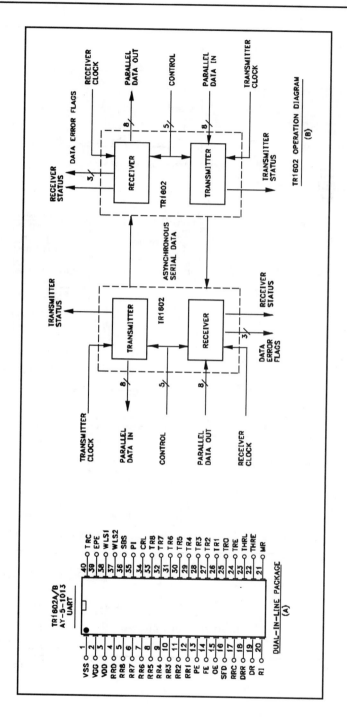

Figure 3.1. The DIP package pinouts and overall block diagram for the TR1602/AY-5-1013 UART.

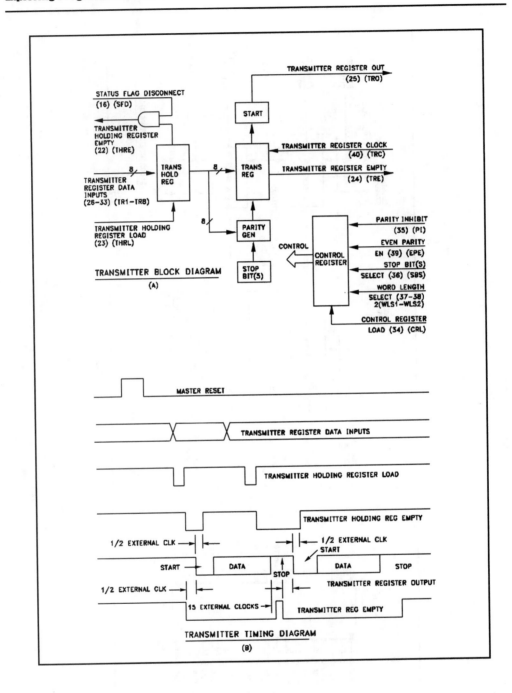

Figure 3.2. Function and timing diagrams for the transmitter section of the TR1602/AY-5-1013 UART.

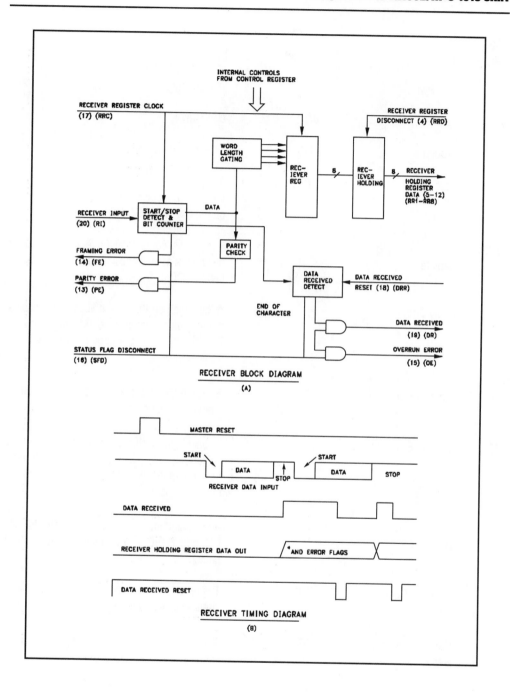

Figure 3.3. Function and timing diagrams for the receiver section of the TR1602/AY-5-1013 UART.

Pin No	Name	Symbol	Function
1	Pos. Pwr	VSS	+5V supply
2	Neg. Pwr	VGG	-12V supply
3	Ground	VDD	Power supply common
4	Rcvr. Reg. Disconnect	RRD	A logic high applied to this input disconnects the disconnects the receiver-holding register output from the RR8-RR1 data outputs.
5-12	Receiver Holding Register Data	RR8-RR1	The parallel contents of the receiver-holding register appear on these lines if a logic low is applied to RRD.
13	Parity Error	PE	A logic high on this line indicates the received parity does not compare to that programmed to Pin 39. This output is updated with each data transfer cycle.
14	Framing Error	FE	A logic high on this line indicates the received character has no valid stop bit.
15	Overrun Error	OE	A logic high on this line indicates the data-received flag (Pin 19) was not reset before the next character was transferred the receiver-holding register.
16	Status Flags Disconnect	SFD	A logic high input disconnects the PE, FE, OE, DR, and THRE, allowing them to be bus connected.
17	Receiver Register Clock	RRC	The receiver clock frequency is 16 times the desired receiver shift rate.
18	Data Received Reset	DRR	A logic low applied to this line resets the DR line.

Table 3.1. *(Continued on next page.)* **TR1602 pin definitions.**

Pin No	Name	Symbol	Function
19	Data Received	DR	A logic high indicates an entire character has been received and transferred to the receiver-holding register.
20	Receiver Input	RI	Serial data received on this line enters the receiver register at a point determined by the character length, parity, and number of stop bits. Must be high when data is not being received.
21	Master Reset	MR	A logic high on this input clears the logic. The transmitter and receiver registers, the receiver-holding register, FE, OE, PE, and DRR are reset. TRO, THRE and TRE are set to the logic high.
22	Transmitter Holding Register Empty	THRE	A logic high on this line indicates the transmitter holding register has transferred its contents to the transmitter register and may be loaded with a new character.
23	Transmitter Holding Register Load	THRL	A logic low applied to this line enters a character into the transmitter-holding register. A low/high transition transfers the character into the transmitter register, unless it is in the process of transmitting a character. In this event, the transfer takes place on initiation of the serial transmission of the new character.
24	Transmitter Register Empty	TRE	A logic high on this line indicates that the transmitter register has completed serial transmission of a full character. It remains in this state until transmission start of the next character.
25	Transmitter Register Output	TRO	The contents of the transmitter register are serially shifted out on this line. The line remains high while no data transfer is taking place. Start of transmission is defined as the high-to-low transition of the start bit.

Pin No	Name	Symbol	Function
26-33	Transmitter Register Data Inputs	TR1-TR8	The character to be transmitted is loaded into the transmitter-holding register with a low-to-high transition of the THRL strobe.
34	Control Register Load	CRL	A logic high loads the control register with the control bits (WLS1, WLS2, EPE, PI, SBS). May be strobed or hardwired high.
35	Parity Inhibit	PI	A logic high inhibits the parity generation; the PE output, Pin 13, is clamped to logic low.
36	Stop Bit(s) Select	SBS	This input determines the stop bits. Logic high selects two; logic low selects one stop bit. When a five bit character is programmed 1.5 stop bits are selected.
37-38	Word Length Select	WLS2-WLS1	These two lines select the character length as follows:
39	Even Parity Enable	EPE	Logic high selects even parity; low selects odd parity.
40	Transmitter Register Clock	TRC	The transmitter clock frequency is 16 times the desired transmitter shift rate.

WLS1	WLS2	Word Length
Low	Low	5 bits
Low	High	6 bits
High	Low	7 bits
High	High	8 bits

Table 3.1. *(Continued from previous two pages.)*

Two versions of the device are available: the TR1602/AY-5-1013A and TR1602B. The maximum frequency for these is 320 kHz. Three versions of the TR1602B, B-03, B-04, and B-05 offer frequencies of 480, 640, and 800 kHz, respectively. Time delays are in the range of 200 to 500 nS, with a 20-nS hold time and zero set time.

CONTROL DEFINITION								
CONTROL WORD						CHARACTER FORMAT		
W L S 2	W L S 1	P I	E P E	S B S	START BIT	DATA BITS	PARITY BIT	STOP BITS
0	0	0	0	0	1	5	ODD	1
0	0	0	0	1	1	5	ODD	1.5
0	0	0	1	0	1	5	EVEN	1
0	0	0	1	1	1	5	EVEV	1.5
0	0	1	X	0	1	5	NONE	1
0	0	1	X	1	1	5	NONE	1.5
0	1	0	0	0	1	6	ODD	1
0	1	0	0	1	1	6	ODD	2
0	1	0	1	0	1	6	EVEN	1
0	1	0	1	1	1	6	EVEN	2
0	1	1	X	0	1	6	NONE	1
0	1	1	X	1	1	6	NONE	2
1	0	0	0	0	1	7	ODD	1
1	0	0	0	1	1	7	ODD	2
1	0	0	1	0	1	7	EVEV	1
1	0	0	1	1	1	7	EVEV	2
1	0	1	X	0	1	7	NONE	1
1	0	1	X	1	1	7	NONE	2
1	1	0	0	0	1	8	ODD	1
1	1	0	0	1	1	8	ODD	2
1	1	0	1	0	1	8	EVEN	1
1	1	0	1	1	1	8	EVEN	2
1	1	1	X	0	1	8	NONE	1
1	1	1	X	1	1	8	NONE	2

Table 3.2. TR1602 configuration control word definitions.

Asynchronous Data Transfer With the TR1602/AY-5-1013

For the transfer of digital data over long distances, a serial mode is generally preferred. Two transmission modes are possible: synchronous and asynchronous. The two differ in how synchronism is maintained between the transmitter and the receiver.

Synchronous transmission is highly structured, with synchronized timing at each end of the data link. To accomplish this, a synchronization signal is transmitted at the beginning of the transmission. A specified clock transition must also be transmitted along with the data to define the data start. Each data bit must follow contiguously after the sync word, as one data bit is associated with each clock period.

The TR1602/AY-5-1013 provides asynchronous transmission only. This mode doesn't require the inclusion of a clock signal with the data. Instead, the data bits are grouped into a word format—generally from five to eight bits in length—with synchronizing start and stop information inserted. The start element is a single logic 0 (space) inserted at the front of each data word. The stop is a logic 1 (mark) added at the end.

The stop is maintained until the next data word is ready for transmission. There is no upper limit to the stop time, but a minimum of one or two data-bit intervals exists. This is selectable with the TR1602/AY-5-1013. The negative-going transition of the start element defines the location of the data bits in each character transmitted. The clock source at the receiver is reset by this transition, which is used to locate the center of each data bit.

This feature provides a significant degree of latitude in receiver timing. The receiver uses a 16X clock for timing. Assuming a symmetrical square wave for the clock, the center of the data bit is defined as 7½ clock counts. Given a perfect clock, the center will always be found within ± 3.125%, giving a margin of 46.875%. In practice, a clock frequency of 9,600 Hz is common.

The transmission rate is measured in bauds, defined as the reciprocal of the shortest signal element. This is typically one data-bit interval. The inclusion of the stop bit causes the baud rate to differ from the bit rate.

The major advantage of asynchronous transmission is that a clock signal need not be transmitted with the data. Another is that the data need not be contiguous in time, but is transmitted as it becomes available. This is particularly important when transmitting from a keyboard. The primary disadvantage of asynchronous transmission is that a large portion of the bandwidth must be used for the synchronizing information.

Transmission over a simple twisted-wire pair can be accomplished at baud rates of 10 kHz or higher. How high depends on the line length and the driver and receiver used. Another limiting factor is the allowable distortion. Distortion is generally created by frequency jitter and offset in the clock source and transmission channel noise. Electronic signal generators can be held to very low tolerances, while electromechanical devices, such as the teletype, may exhibit distortions as high as 20%.

TR1602/AY-5-1013 Operation

The device is capable of full-duplex (simultaneous transmission and reception) or half-duplex operation. The transmitter portion assembles the parallel input into a serial asynchronous data system. Control options permit sizing data characters into lengths of five, six, seven, or eight bits. Even or odd parity may be assigned. The baud rate can be set anywhere from DC to 12 kHz by providing a transmitter clock of 16 times the desired rate.

The receiver section restructures the asynchronous input into a parallel data character. It does this by searching for the start bit, finding the center of each data bit, and outputting the character in a parallel format stripped of the start, parity, and stop bits. Three error flags are provided to indicate if there is an error in the parity, if a valid stop bit was not decoded, or if the last character wasn't unloaded by the external device before the next character was received. As with the transmitter, the receiver clock is set at 16 times the transmission baud rate.

Both transmitter and receiver feature double buffering. In this way, one complete character is always available internally. This facilitates exchange with the external devices.

The device DIP package pinouts and simplified block diagrams for the transmitter and receiver sections are shown in Figure 3.1. It will be helpful to relate the functions of Figures 3.2 and 3.3 to the definitions of Table 3.1.

Figure 3.2 further defines the block diagram of the transmitter portion of the device. Data may be loaded into the transmitter holding register whenever the transmitter holding register empty (THRE) is at a logic 1, indicating that the holding register is empty. The data is loaded in by strobing the transmitting holding register load (THRL) to logic 0. The data is automatically transferred to the transmit register as soon as the register becomes empty. The required start, stop, and parity bits are then added and serial transmission begins.

The number of stop bits and parity type is specified by the control register. The control line status is strobed in by a low-to-high transition of the control register load (CRL) line. The various control options are defined in Table 3.2.

A master reset (MR) input sets the transmitter to the idle state when its line is at a logic 1. A status flag disconnect (SFD) is provided. When this line is at logic 1 level, the THRE line is disabled and goes to a high impedance. Thus, this output from several devices may be tied to the same bus.

In operation, the master reset should be strobed immediately following power on. The external devices can then set the transmitter register data inputs. Upon receipt of a load pulse, the data is transferred to the transmitter register where the start, parity, and stop bits are added and transmission started.

The timing diagram included with the diagram is helpful in understanding the sequence of operations. Particularly helpful is the illustration of the framing of the asynchronous serial data for transmission.

After power on, the master reset should be strobed to set the circuits to the idle state. The external device can then set the transmitter register data inputs to the desired value, and after the data inputs are stable, the load pulse is applied. The data is then automatically transferred to the transmitter register, where the start, stop, and parity bits are added and the transmission is started. The process is repeated for each subsequent character as it becomes available. The only timing requirement for the external device is that the data inputs be stable during the load pulse and for 20 nS following.

Figure 3.3 is a similar expansion of the block diagram of the receiver portion of the device. Serial input data is provided to the receiver input (RI). A start bit detect circuit continually searches for a logic 1-to-0 transition while in the idle state. When this transition is detected, a counter is initiated that begins a search for the center of the start bit. If the input is still a logic 0 at the center, a valid start bit is presumed. The counter then continues in its count to locate the centers of all subsequent data and stop bits. This procedure guards against mistaking noise or other erroneous input for valid data.

Like the transmitter, the receiver is under control of the control register. A word-length gating circuit adjusts the length of the receiver register to match the length of the incoming data word. A parity-check circuit verifies the parity, or sets an error flag to logic 1 if needed. Another circuit checks the first stop bit of each character. If the stop bit is not a logic 1, the framing error (FE) flag is set to logic 1.

As each received character is transferred to the holding register, the data-received (DR) line is set to logic 1 as a signal to the external device. When this device reads the data, it should strobe the data-received reset (DRR) line to a logic 0 to reset the DR line. If the reset is not done before a new character is transferred, it will be lost. In this event, the overrun error (OE) line will be set to a logic 1.

Figure 3.3 also shows the relative timing of the receiver signals. A master reset strobe places the unit in the idle mode, and the receiver begins a search for the first start bit. After a complete character has been decoded, the data output and error flags are set to the correct level and the data-received line is set to logic level one. The data outputs are set to the proper level one-half clock period before the DR and error flags, which are set in the center of the first stop bit. The data-received reset pulse resets the DR line to a logic 0.

As with the transmitter portion, the timing for the receiver functions is 16 times the transmission baud rate.

Figure 3.4 provides a description of the switching characteristics with illustrative timing diagrams.

Figure 3.5 illustrates a TR1602/AY-5-1013 modem application with a computer communications controller. In this example, the modem is sourcing serial data via an RS-232 data link to the UART's receiver section, where it is restructured as an eight-bit data word for transmission in parallel to the communications controller. The data is outputted from the computer on an eight-bit data bus to the transmitter section, which restructures it to the required transmission mode and sends it out in serial format to the RS-232 data link for transmission to the modem.

SWITCHING CHARACTERISTICS
(Vss = Vcc = 5V, VGG = −12V, TA = 25°, CLoad = 20pf plus one TTL Load)

SYM	PARAMETER	MIN	MAX	CONDITIONS
fclock	Clock Frequency	DC	320KHz*	Vss=4.75V
tPW	Pulse Widths			(Figs 1 & 2)
	CRL	200nS		
	THRL	200nS		
	DRR	200nS		
	MR	500nS		
tc	Coincidence Time	200nS		(Figs 1 & 2)
thold	Hold Time	20nS		(Figs 1 & 2)
tset	Set Time	0		(Figs 1 & 2)
	Output Propagation Delays			(Fig. 3)
tpd0	To Low State		500nS	CL 20pf, plus one TTL load
tpd1	To High State		500nS	(Fig. 3)
	Capacitance			CL 20pf, plus one TTL load
Cin	Inputs		20pf	f = 1MHz, Vin = 5V
Co	Outputs		20pf	f = 1MHz, Vin = 5V

*fmax for TR1602A or B

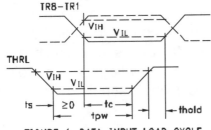

FIGURE 1. DATA INPUT LOAD CYCLE

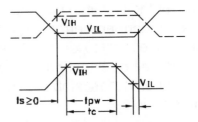

FIGURE 2. CONTROL REGISTER LOAD CYCLE

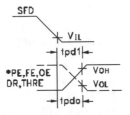

FIGURE 3. STATUS FLAG OUTPUT DELAYS
*Outputs of PE,FE,OE,DR,THRE are disconnected at transition of SFD from VIL to VIH.

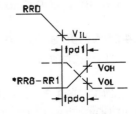

FIGURE 4. DATA OUTPUT DELAYS
*RR8–RR1 are disconnected at transition of RRD from VIL to VIH.

Figure 3.4. Switching characteristics with waveforms of the TR1602/AY-5-1013 UART.

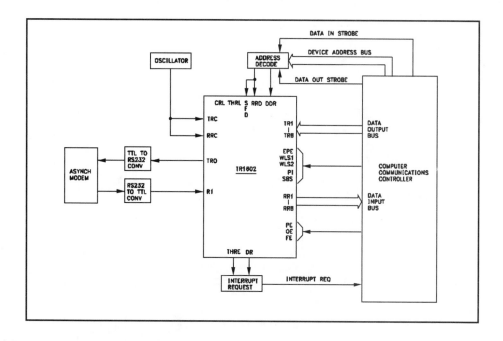

Figure 3.5. A modem application of the TR1602 UART with a computer communications controller.

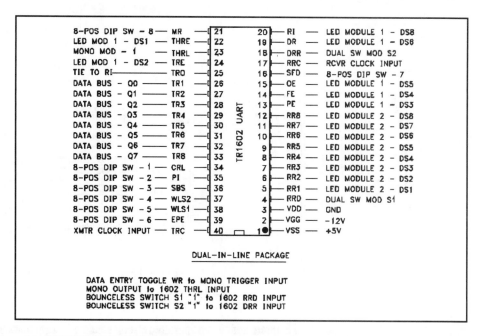

Figure 3.7. TR1602/AY-5-1013 wiring plan for bench-top operation.

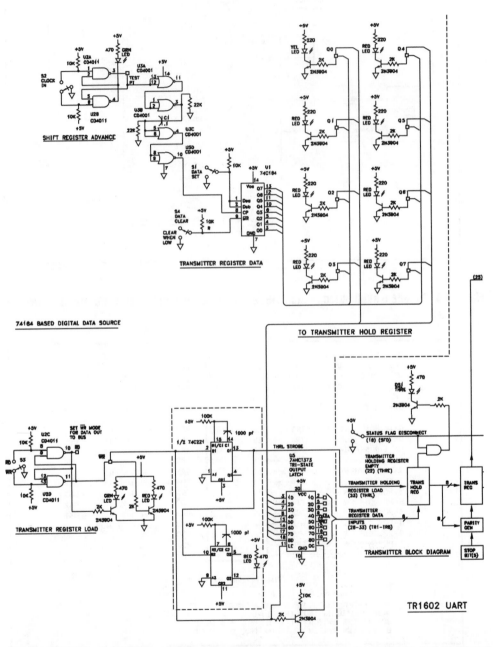

Figure 3.6. *(Continued on next page.)* **The complete bench-top operating circuit schematic. Schematics of Chapter 1 modules are included for ease of understanding.**

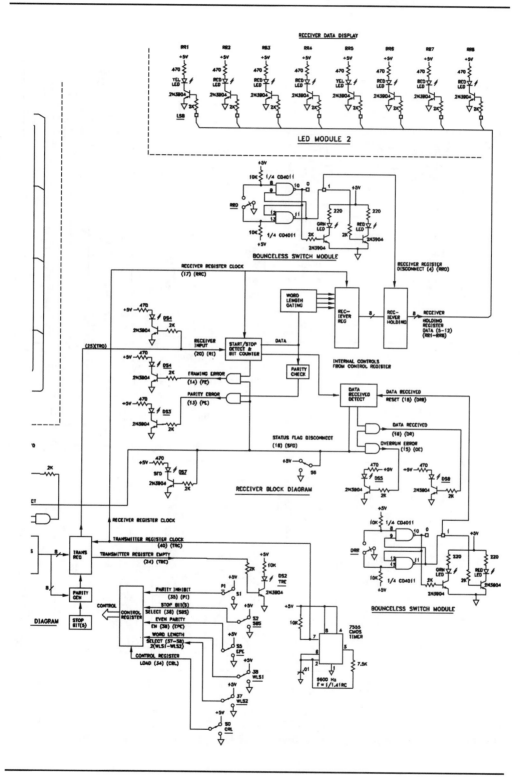

RECEIVER DATA DISPLAY

LED MODULE 2

BOUNCELESS SWITCH MODULE

RECEIVER BLOCK DIAGRAM

BOUNCELESS SWITCH MODULE

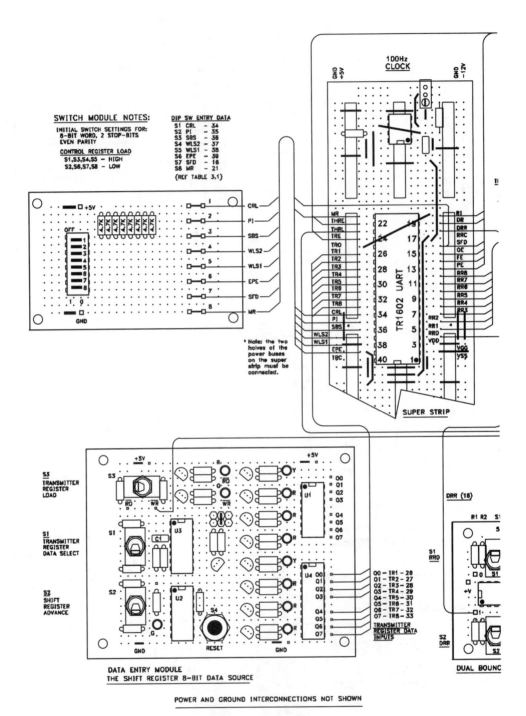

Figure 3.8. *(Continued on next page.)* **The author's bench-top operating circuit module layout.**

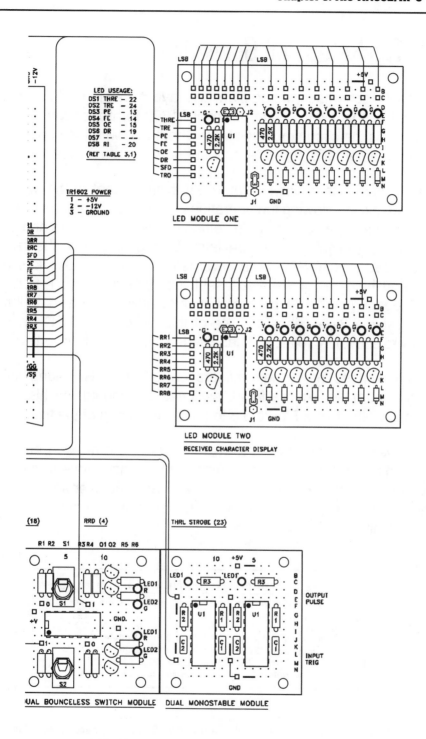

LED MODULE ONE

LED MODULE TWO
RECEIVED CHARACTER DISPLAY

DUAL BOUNCELESS SWITCH MODULE   DUAL MONOSTABLE MODULE

Bench-Top Setup and Operation

We will use the schematic, wiring plan, and layout drawings of *Figures 3.6, 3.7*, and *3.8*, respectively, to set up the TR1602 for bench-top operation. Figure 3.8 uses the circuit modules from Chapter 1, with the super strip or a similar experimenter's equivalent for support of the TR1602.

The overall circuit schematic of Figure 3.6 includes that of the data-entry module as an aid in following the operations. Note the operation of switches S1 and S2 in the upper left corner in setting up the pattern of 1s and 0s that are input to U4, the 74HCT373 output latch. During this phase, the transmitter register load switch, S3, at the lower left, is in the RD position. When ready to send the data out to the bus, set this switch momentarily to the WR position, triggering the THRL monostable and passing the bus data through the latch. *Table 3.3* describes the use of this switch. The LED circuits of Module 1 are shown in detail, with labels identifying their function.

Figure 3.7 describes the wiring configuration for the UART, along with the data entry and the two external toggle switches. We should use this drawing in conjunction with that of Figure 3.8 to assist our wiring of the test setup.

In normal system usage, there are two devices—one at each end of the transmission line. However, we can run a performance procedure with just a single device in that each half—transmitter and receiver section—is independently operated. All we need do is jumper the transmitter serial output to the receiver serial input.

HOW TO USE THE DATA ENTRY R/W SWITCH WITH THE TR1602 UART

With this IC data entry on the bus is not related to the clock.

1. To write to the Data Bus:
   Ensure the Data Entry R/W switch is in the R position.
   Establish the Data Pattern on the Data Entry Module
2. Toggle the Data Entry R/W switch to the W position.
3. Return the Data Entry R/W switch to the R position.
   A display of the entered data will appear on LED Module 2.

Table 3.3. Using the R/W toggle switch on the data-entry module for entering data on the TR1602 bus for transmission.

As a first step, it's a good idea to get the clock up and running. Though the schematic shows 100 Hz, the actual frequency is not critical. Keeping the frequency low allows us to view operations in slow motion with the display LEDs.

The layout shown is only a suggestion, but if you have constructed the modules as described in Chapter 1, it is a very convenient approach. Pay particular attention to the power connections for UART pins 1, 2, and 3. Note the power to pin 2 is a *negative* 12V. With the two drawings as guides, the entire test arrangement is easily set up and run.

The assembly drawing shows the initial position of the toggle switches employed. *Figure 3.9* is a procedure sheet for the initial setup and operation of the test system. The procedure shown will give us familiarity with the setup. Later, we may wish to vary the conditions to observe the effects of control options. Note that for a properly functioning UART, the three LED error indicators for PE, FE, and OE will never light up.

We may want to follow along with the procedure (Figure 3.9) as we read the remainder of the text here. When we first turn on the power, the eight LEDs on the data-entry module come up in a random state, some bright, others dark. We should press reset switch S4 to clear the shift register to all zeros. We also toggle DIP switch S8, (MR), to reset the UART. The status of the display LEDs is shown on the procedure of Figure 3.9.

The receiver-loading register is disconnected by switch RDD; the RR1 - RR8 LED display, Module 2, should be dark. Module 1 LEDs DS1, DS2, and DS6 are bright. If you are watching DS5 and DS6 at turn on, you might see a momentary flash.

Suppose we start by pressing the data-entry module reset to clear the shift register outputs. Then we might key in an alternating 1010 1010 pattern using the module's transmitter register data and shift register advance switches. Next, we transfer the settings through the 74HCT373 latch to transmitter holding register (THRL) using the RD/WR switch. LED Module 1 DS8 will blink with the 1s and 0s in the transmitted serial data.

BENCH RUN PROCEDURE FOR THE TR1602/AY-5-103 UART

CAUTION: Review Table 3.3 before beginning this procedure to ensure correct useage of the Data Entry and Toggle Switch Modules.
Refer to Figures 3.7 and 3.8 for the bench top wiring setup.

1. Initialize the Bench Top Wiring Setup.
   Setup the DIP switch module as shown. ☐
   MR momentary "1" at power on. ☐
   Verify toggles DRR and RRD at "1" position. ☐

OFF

		LED Module 1 Status:				LED Module 2 Status
1	CRL	DS1	THRE	ON	DS1	RR1 OFF
2	PI	DS2	TRE	ON	DS2	RR2 OFF
3	SBS	DS3	PE	OFF	DS3	RR3 OFF
4	WLS2	DS4	FE	OFF	DS4	RR4 OFF
5	WLS1	DS5	OE	OFF	DS5	RR5 OFF
6	EPE	DS6	DR	OFF	DS6	RR6 OFF
7	SFD	DS7	SFD	OFF	DS7	RR7 OFF
8	MR	DS8	RI	ON*	DS8	RR8 OFF

1 0

DIP SW SETTINGS   ☐   *RI MAY PULSE AT POWER ON   ☐

Verify that toggle switches are positioned as shown in Figure 3.7. ☐

2. Suggested first Transmitter Register Data load. ☐

DATA BUS

7 6 5 4 3 2 1 0	OPERATION
1 0 1 0 1 0 1 0	Bus Data Entry

BUS DATA ENTRY = 1010 1010 ☐

Toggle Data Entry Module switch S3 to pulse THRL.
Observe serial transmission on LED Module 1, DS8 (RI).

LED Module 1 Status:			LED Module 2 Status		
DS1	THRE	ON	DS1	RR1	OFF
DS2	TRE	ON 1	DS2	RR2	ON
DS3	PE	OFF	DS3	RR3	OFF
DS4	FE	OFF	DS4	RR4	ON
DS5	OE	OFF	DS5	RR5	OFF
DS6	DR	ON 3	DS6	RR6	ON
DS7	SFD	OFF	DS7	RR7	OFF
DS8	RI	ON 2	DS8	RR8	ON

NOTE:
Will duplicate the data entered on the bus with toggle RRD at "0" position

☐   ☐

1. OFF while sending data
2. Pulses at data rate while sending.
3. Toggling switch DRR turns off.

3. Continue with new data entries.

Figure 3.9. TR1602/AY-5-1013 UART test setup and operating procedure.

At this time, there should be no change on the RR1 - RR8 LEDs; they remain dark. LEDs DS1 and DS2 (THRE and TRE) are bright, with DS3, DS4, and DS5 (PE, FE, OE) dark. DS6 (DR) is bright.

This is as it should be. RR1 - RR8 are dark because switch S1 on the bounceless switch module, RRD, in the high position disconnects the receiver-holding register content from the output lines.

DS1 bright indicates the transmitter-holding register has transferred its content to the transmitter register and is ready for a new character.

DS2 bright indicates the transmitter register has completed the serial transmission of one full character. DS2 will be momentarily off while the serial transmission is taking place. It will remain on until the start of the next transmission. Because of the slow speed, we will see this LED go dark as the transmission proceeds.

Now suppose we set switch RRD to the low position. LEDs RR1 - RR8 will now duplicate the pattern of the control module LEDs. This feature of the UART is needed when the data transfer is pulsed.

LEDs DS3, DS4, and DS5 are always dark when the UART is functioning correctly. DS3 bright means we have a parity error. DS4 bright means we have a framing error. DS5 bright means we have an overrun error. It's difficult to create parity and framing errors, but we can force an overrun error by toggling the transmitter register load switch twice in a row without toggling DRR. What is needed is to toggle switch DRR between each transmission to reset the data-received flag. This clears DS6, which will remain dark until the next transmission.

It is rewarding to vary the shift-register pattern and observe the faithful reproduction at the receiver outputs. We note that DS8 remains bright while awaiting the next transmission. Why is this? An oscilloscope will provide an even better picture of the transmission, of course. This is especially true if we vary the data to include more than a single high or low in the series sequence, though if we watch DS8 closely we will see the increased on time.

Note that toggling switch S8 resets the UART. We don't need to do this other than at the beginning, as noted on the procedure, but we do need to reset the data-received line (DR) by toggling switch DRR before sending new data.

Table 3.1 defines the functions of those inputs connected to the eight DIP-module switches. It will be instructive to vary these switch positions to observe the effects of setting changes.

References

1. Data sheet, Western Digital, *MOS/LSI TR1602A & TR1602B Asynchronous Receiver/Transmitter.*

2. Application Report #1, Western Digital, *MOS/LSI TR1602A & TR1602B Asynchronous Receiver/Transmitter.*

3. References for Figure 3.1: Data sheet, Western Digital, *MOS/LSI TR1602A & TR1602B Asynchronous Receiver/Transmitter*, DIP pin assignments. Application Report #1, Western Digital, *MOS/LSI TR1602A & TR1602B Asynchronous Receiver/Transmitter,* Figure 5.

4. Reference for Figure 3.2: Application Report #1, Western Digital, *MOS/LSI TR1602A & TR1602B Asynchronous Receiver/Transmitter,* Figures 6 and 7.

5. Reference for Figure 3.3: Ibid. Figures 8 and 9.

6. Reference for Figure 3.4: Data sheet, Western Digital, *MOS/LSI TR1602A & TR1602B Asynchronous Receiver/Transmitter*, Table: Switching Characteristics, Figures 1-4.

7. Reference for Figure 3.5: Application Report #1, Western Digital, *MOS/LSI TR1602A & TR1602B Asynchronous Receiver/Transmitter,* Figure 12.

8. Reference for Table 3.1: Data sheet, Western Digital, *MOS/LSI TR1602A & TR1602B Asynchronous Receiver/Transmitter*, Table: Pin Definitions.

9. Reference for Table 3.2: Application Report #1, Western Digital, *MOS/LSI TR1602A & TR1602B Asynchronous Receiver/Transmitter,* Table 1.

Chapter 4
The SY6522/SY6522A VIA

Introduction

Back in 1976, I purchased a KIM-1 single-board computer, based on the 6502 microprocessor. The manufacturer was MOS Technology. A 6520 interface adapter was provided for use with peripherals. It contained 16 bidirectional ports, but the KIM-1 used only 15, which to my mind was a significant shortcoming for this otherwise truly wonderful little computer. One port, PB6, was diverted for another use. This resulted in an awkward situation for many applications. The SY6520, the Motorola MC6820 and MC6821, are the devices described in Chapter 5.

Sometime later I upgraded to the Synertek SYM-1, a mostly KIM-1 compatible. This computer employed the SY6522[1-6], a more versatile device—and considerably more complex and challenging in its application.

Let's begin with a brief summary of just what role an interface adapter plays in the operation of a microprocessor. The major requirement is the provision of communication between the microprocessor and the external world—that is, peripheral devices. For the 6522, 16 bidirectional I/O ports are organized as two eight-bit ports, designated A and B. In addition to the I/O provision, there are two 16-bit programmable timer/counters, a serial-to-parallel/parallel-to-serial shift register, and input data latching on the ports.

Each port can be individually programmed as an input or output. Two peripheral I/O lines (PB6 and PB7) have functions controlled from the interval timers (T1 and T2). This capability permits the generation of programmable one-shots and frequency square waves. It also enables the counting of externally generated pulses.

Registers are included for peripheral data, peripheral port data direction, peripheral and auxiliary control, data shifting, interrupt flags, and interrupt enable.

The 6522 is a 1-MHz device; the 6522A operates at clock speeds of up to 2 MHz. These relate to data read and write capabilities only, as shown in *Figure 4.2*. Peripheral timing characteristics and related diagrams are provided in *Figure 4.3*. The two device types are identical and no distinction other than timing differentiates the two. These features will be fully described in the sections that follow.

Theory of Operation

Let's begin with *Figure 4.1*. Here we see the package pinout and a block diagram of the major internal functions. The 6522 is provided in a 40-pin dual-in-line package (DIP). The ground connection is made to pin 1, the +5V dc power to pin 20. No negative voltage power is required.

The block diagram shows us a buffered bidirectional data bus; the read/write control (R//W) is seen in the chip access control block. The block under interrupt control contains the flag register (IFR) and the interrupt enable register (IER). The function control block also contains two registers: the peripheral control register (PCR) and the auxiliary control register (ACR).

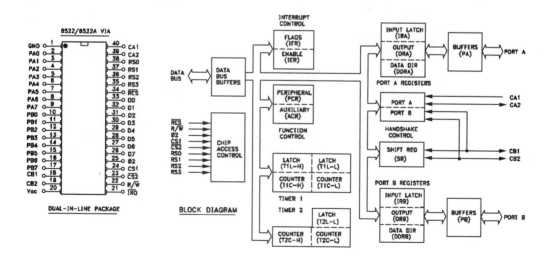

Figure 4.1. The 6522 VIA DIP package and block diagram.

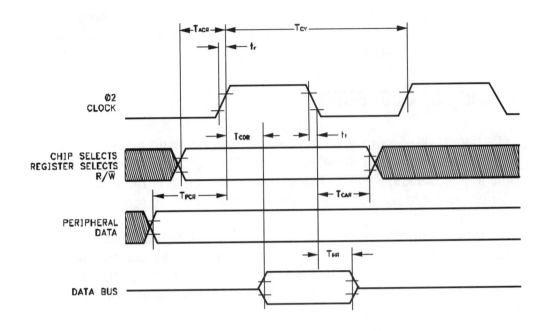

Symbol	Parameter	6522		6522A		Unit
		Min	Max	Min	Max	
T_{CY}	Cycle Time	1	50	0.5	50	uS
T_{ACR}	Address Set-Up Time	180	–	90	–	nS
T_{CAR}	Address Hold Time	0	–	0	–	nS
T_{PCR}	Peripheral Data Set-Up Time	300	–	300	–	nS
T_{CDR}	Data Bus Delay Time	–	395	–	200	nS
T_{HR}	Data Bus Hold Time	10	–	10	–	nS

Note: tr, tf = 10 to 30 nS

READ TIMING CHARACTERISTICS

Figure 4.2A. The 6522 data read timing characteristics.

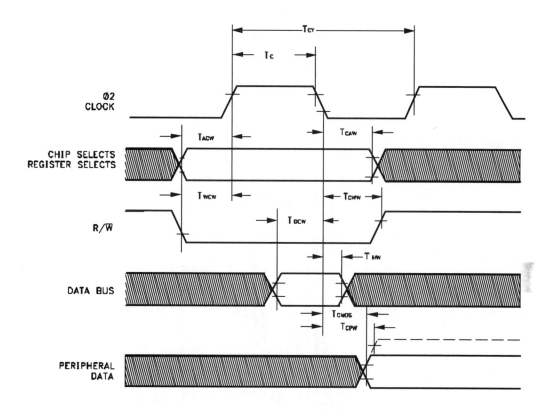

Symbol	Parameter	6522		6522A		Unit
		Min	Max	Min	Max	
T_{CY}	Cycle Time	1	50	0.50	50	uS
T_{D}	Ø2 Pulse Width	0.47	–	0.25	25	uS
T_{ACW}	Address Set-Up Time	180	–	90	–	nS
T_{CAW}	Address Hold Time	0	–	0	–	nS
T_{WCW}	R/$\overline{W}$ Set-Up Time	180	–	90	–	nS
T_{CWW}	R/$\overline{W}$ Hold Time	0	–	0	–	nS
T_{DCW}	Data Bus Set-Up Time	300	–	150	–	nS
T_{HW}	Data Bus Hold Time	10	–	10	–	nS
T_{CPW}	Peripheral Data Delay Time	–	1.0	–	1.0	uS
T_{CMOS}	Peripheral Data Delay Time to CMOS Levels	–	2.0	–	2.0	uS

Note: tr, tf = 10 to 30 nS

WRITE TIMING CHARACTERISTICS

Figure 4.2B. The 6522 data write timing characteristics.

PERIPHERAL INTERFACE CHARACTERISTICS

Symbol	Characteristic	Min	Max	Unit	Figure*
tr,tf	Rise and Fall Time for CA1,CB1,CA2,and CB2 Input signals	—	1.0	uS	—
tCA2	Delay Time, Clock Negative Transition to CA2 Negative Transition (read handshake or pulse mode)	—	1.0	uS	5a,5b
tRS1	Delay Time, Clock Negative Transition to CA2 Positive Transition (pulse mode)	—	1.0	uS	5a
tRS2	Delay Time, CA1 Active Transition to CA2 Positive Transition (handshake mode)	—	2.0	uS	5b
tWHS	Delay Time, Clock Positive Transition to CA2 or CB2 Negative Transition (write handshake)	—	1.0	uS	5c,5d
tDS	Delay Time, Peripheral Data Valid to CB2 Negative Transition	0	1.5	uS	5c,5d
tRS3	Delay Time, Clock Positive Transition to CA2 or CB2 Positive Transition (pulse mode)	—	1.0	uS	5c
tRS4	Delay Time, CA1 or CB1 Active Transition to CA2 or CB2 Positive Transition (handshake mode)	—	2.0	uS	5d
tIL	Setup Time, Peripheral Data Valid to CA1 or CB1 Active Transition (input latching)	300	—	nS	5e
tSR1	Shift Out Delay Time — Time from Ø2 Falling Edge to CB2 Data Out	—	300	nS	5f
tSR2	Shift In Setup Time — Time from CB2 Data In to /Ø2 Rising Edge	300	—	uS	5g
tIPW	Pulse Width — PB6 Input Pulse	2	—	uS	5i
tICW	Pulse Width —CB1 Input Clock	2	—	uS	5h
tIPS	Pulse Spacing — PB6 Input Pulse	2	—	uS	5i
tICS	Pulse Spacing — CB1 Input Pulse	2	—	uS	5h

*Figure numbers refer to those shown in the reference Data Manual.

Figure 4.3. *(Continued on next page.)* **The 6522 peripheral interface characteristics and related timing diagrams.**

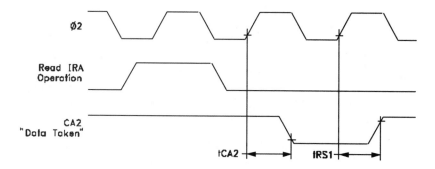

Fig. 5a. CA2 Timing for Read Handshake, Pulse Mode.

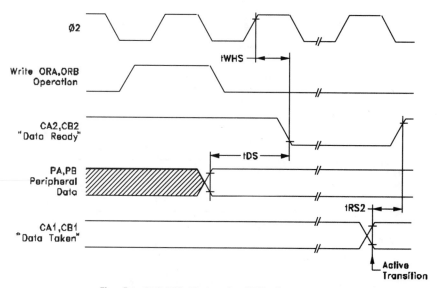

Fig. 5d. CA2,CB2 Timing for Write Handshake, Handshake Mode.

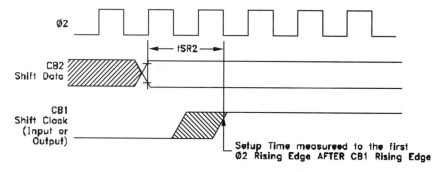

Fig. 5g. Timing for Shift In with Internal or External Clocking.

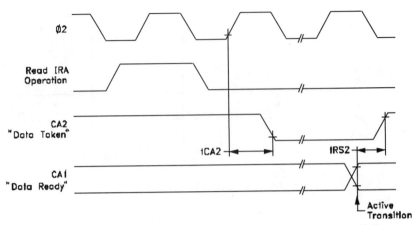

Fig. 5b. CA2 Timing for Read Handshake, Handshake Mode.

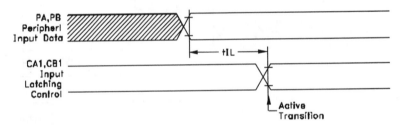

Fig. 5e. Peripheral Data Input Latching Timing.

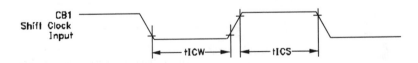

Fig. 5h. External Shift Clock Timing.

Figure 4.3. *(Continued from previous two pages.)*

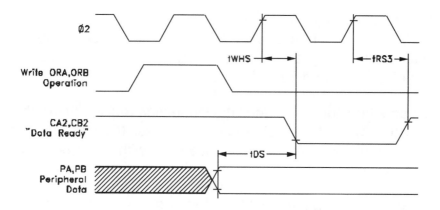

Fig. 5c. CA2,CB2 Timing for Write Handshake, Pulse Mode.

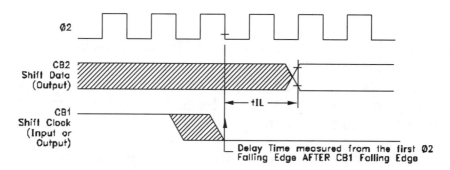

Fig. 5f. Timing for Shift Out with Internal or External Shift Clocking.

Fig. 5i. Pulse Count Input Timing.

The Timer 1 block contains four registers: the low-order latch and counter, T1L-L and T1L-H, and T1C-L and T1C-H. The Timer 2 block contains one less register. Included are the low-order latch, T2L-L, and low- and high-order counters, T2C-L and T2C-H.

The Port A peripheral block includes the input-latch register (IRA), the output register (ORA), and the data-direction register (DDRA). The Port B peripheral block provides the same functions as with the A port. Each is buffered as shown.

A handshake control feature exists for each port peripheral block. Read and write handshake controls (CA1 and CA2) are provided for Port A. Write handshake controls only (CB1 and CB2) exist for Port B. These are used also for the eight-bit shift register (SR).

Table 4.1 provides a definition of the symbols used with the device pinouts and a summary of the function block features shown in Figure 4.1.

Table 4.2 defines the register coding and the functions they perform. Observe the organization of Table 4.2 with registers shown in groups of four in relation to their logic coding. We will need to have these definitions well in mind as we continue our study of the device and the operations we will conduct with them on the bench.

Table 4.3 is a listing of each register with the figure number in which its function is defined.

The chip selects, CS1/CS2, permit us to designate a specific 6522 from one of several in a system. Since we are working with a single 6522, we will simply set CS1 to a logic 1 and CS2 to a logic 0, +5V dc and ground, respectively.

The 6522 Registers

When we look at Table 4.2 we see that the first four registers, 0-3, are assigned to input/output control and data direction needs. A second group of six, numbers 4-9, are assigned to the low-order and high-order data

PIN No.	SYMBOL	USEAGE
1	GND	The circuit ground connection
2-9	PA0-PA7	Programmable Input/Output pins under control of Data Direction Register A (DDRA). A logic 0 in DDRA sets the pin as an Input. A logic 1 in DDRA sets the pin as an Output. (Ref. Figure 4.5) Output polarity controlled by Output Register A (ORA). A logic 1 in the register causes the output to go high. Reading Port A puts the content of the Input register (IRA) to be transferred to the Data Bus. With input latching disabled IRA reflects the levels on the PA pins at the time latching occurs (via CA1) (Ref Figure 4.
10-17	PB0-PB7	Programmable Input/Output pins under control of Data Direction Register B (DDRB). A logic 0 in DDRB sets the pin as an Input. A logic 1 in DDRB sets the pin as an Output. (Ref. Figure 4.5) Output polarity controlled by Output Register B (ORB). A logic 1 in the register causes the output to go high. Reading Port B puts the content of the Input register (IRB) on the Data Bus. When reading IRB the bit stored in the output register, ORB, is the bit sensed. Reading IRB reads the "1" or "0" level actually programmed. The Input Latching modes are selected by the Auxiliary Control Register (ACR)
18,19	CB1,CB2	These peripheral control lines act as Interrupt Inputs or as handshake outputs. Each line controls an Interrupt enable bit. In addition for this port, these lines act as a serial port under control of the Shift Register. These lines represent one standard TTL load in the Input mode and will drive one standard TTL load in the output mode. They cannot drive Darlington transistor loads.
20	Vcc	Positive DC power, +5V
21	IRQ	This output line goes low whenever an internal interrupt flag is set and the corresponding interrupt enable bit is a logic "1". This output is open drain to allow the request signal to be "wire-or'd" with other equivalent signals in the system.
22	R/W	This input determines the direction of data transfers to and from the 6522. When the line is low data is written into the selected register. When the line is high and the chip selected data is read out of the 6522.
24,23	CS1,CS2	Chip Select Inputs. The selected register will be accessed when CS1 is high and CS2 is low. This feature provides for more a single 6522 in a system.
25	ø2	Phase two clock input. Phase 2 is in reference to the second phase of the 2-phase processor clock. In a microprossor application this clock triggers all data transfers between the system processor and the 6522.
33-26	D0-D7	These eight Data Bus lines are bi-directional. During Read cycles the content of the selected 6522 register are placed on the Data Bus. During Write cycles these lines are high Impedance Inputs and Data is transferred into the selected register. When unselected the lines are high impedance.
34	RES	Reset. When low the reset clears all internal registers to logic "0" with the exception of the T1 and T2 latches and counters and the Shift Register. All peripheral interface lines are set as Inputs. Timers and the Shift Register is disabled and Interrupting from the chip.
35-38	RS3-RS0	These four lines are register selects. Their selection functions are coded in Table 4.2.
40,39	CA1,CA2	These peripheral control lines act as Interrupt Inputs or as handshake outputs. Each line controls an Interrupt enable bit. In addition, CA1 controls the latching of Data on Port A input lines. CA1 is a high impedance input only. CA2 represents one standard TTL load in the Input mode and will drive one standard TTL load in the Output mode. They cannot drive Darlington transistor loads.

Table 4.1. The 6522 VIA DIP pin assignments and a summary of the device functions.

Register	RS Coding				Register	Description	
Number	RS3	RS2	RS1	RS0	Desig'n	Write	Read
0	0	0	0	0	ORB/IRB	Output Register "B"	Input Register "B"
1	0	0	0	1	ORA/IRA	Output Register "A"	Input Register "A"
2	0	0	1	0	DDRB	Data Direction Register "B"	
3	0	0	1	1	DDRA	Data Direction Register "A"	
4	0	1	0	0	T1C-L	T1 Low-Order Latches	T1 Low Order Counter
5	0	1	0	1	T1C-H	T1 High-Order Counter	
6	0	1	1	0	T1L-L	T1 Low-Order Latches	
7	0	1	1	1	T1L-H	T1 High Order Latches	
8	1	0	0	0	T2C-L	T2 Low Order Latches	T2 Low Order Counter
9	1	0	0	1	T2C-H	T2 High-Order Counter	
10	1	0	1	0	SR	Shift Register	
11	1	0	1	1	ACR	Auxiliary Control Register	
12	1	1	0	0	PCR	Peripheral Control Register	
13	1	1	0	1	IFR	Interrupt Flag Register	
14	1	1	1	0	IER	Interrupt Enable Register	
15	1	1	1	1	ORA/IRA	Same as Reg 1 except no "Handshake"	

Table 4.2. The 6522 VIA register number assignments, coding, and designations.

REG NO.	REG NAME	CHAPTER FIGURE
0	ORB/IRB	4.4
1	ORA/IRA	4.5
2	DDRB	4.6
3	DDRA	4.6
4	Timer 1 Low Order Counter	4.10
5	Timer 1 High Order Counter	4.11
6	Timer 1 Low Order Latches	4.12
7	Timer 1 High Order Latches	4.13
8	Timer 2 Low Order Counter	4.15
9	Timer 2 High Order Counter	4.16
10	Shift Register	4.18
11	Auxiliary Control	4.19
12	Peripheral Control	4.8
13	Interrupt Flag	4.22
14	Interrupt Enable	4.23
15	ORA/IRA less handshake	---

Table 4.3. The 6522 VIA register numbers with the related chapter figure numbers.

latches and counters. Register 10 is the shift register. Register 11 is an auxiliary control register and register 12 is a peripheral control register. Registers 13 and 14 are for interrupt functions, and Register 15 is the same as register 1 minus the "handshake" capability.

Port A and B Registers

Each peripheral port is provided with an eight-bit data-direction register (DDRA, DDRB) for specifying the input/output status of its associated port. Each pin can be individually set. For instance, a "0" in a bit of the DDR causes the corresponding pin to function as an input, while a "1" causes the pin to act as an output. Think of "0" as Low (data flows in) and "1" as high (data flows out).

Each peripheral pin is also controlled by a bit in the output register (ORA, ORB) and an input register (IRA, IRB). A pin programmed as an output is further controlled by a bit in the output register. A "1" in this register causes the output to go high, while a "0" causes it to go low. Data may be written into output register bits corresponding to those programmed to be inputs. In this event, the output is unaffected.

Reading a peripheral port causes the content of the input registers IRA, IRB to be transferred onto the data bus. When the input latching is disabled, register IRA will always reflect the levels on the PA pins. With input latching enabled, IRA will reflect the levels on the PA pins at the time the latch took effect via CA1.

The IRB register operates in a manner similar to the IRA register, with a difference for pins programmed as outputs. When reading IRA, the level on the pin determines whether a "0" or a "1" is sensed. When reading IRB, however, the bit stored in the output register ORB is the bit sensed. The reason for the difference is possible loading effects on Port A.

A reading error with Port A may occur with heavy loads that pull a "1" level down or lift a "0" level up, giving a false reading as a result. Reading IRB will always yield the intended "1" or "0" actually programmed.

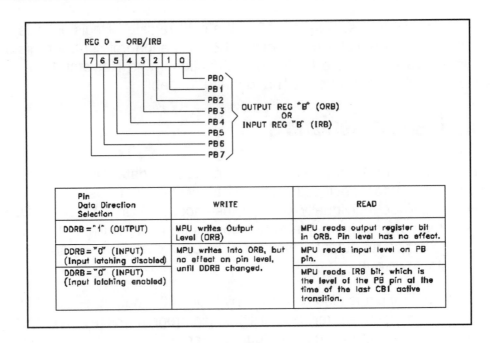

Figure 4.4. Port B input and output (ORB/IRB) register formats.

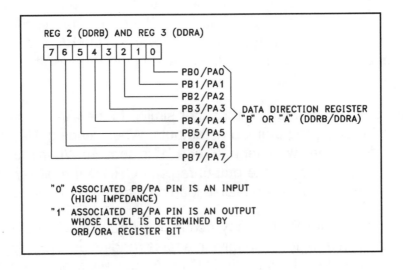

Figure 4.6. Data-direction register formats for Port A (DDRA) and B (DDRB).

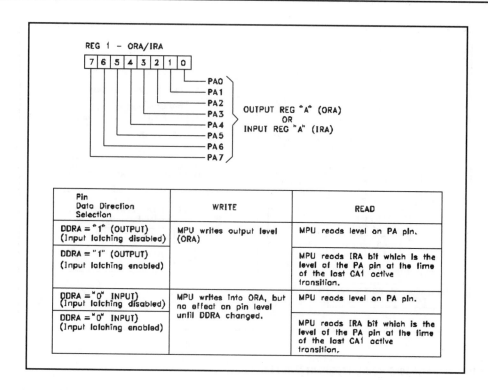

Figure 4.5. Port A input and output (ORA/IRA) register formats.

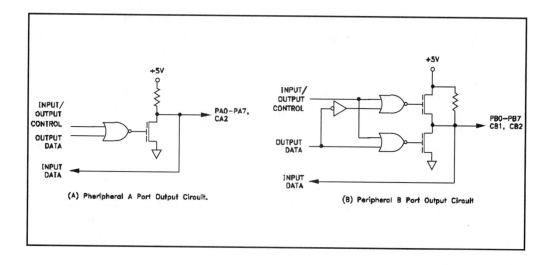

Figure 4.7. Peripheral port output circuitry for Port A and Port B.

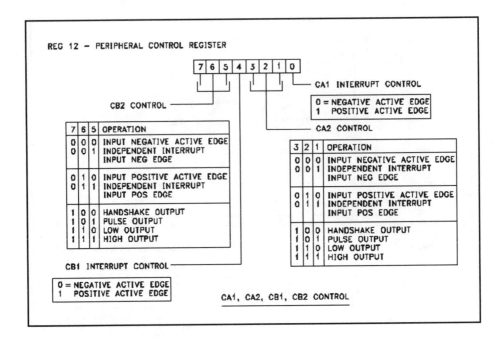

Figure 4.8. Peripheral control register (PCR) format.

Port A data lines represent one standard TTL load in the input mode and will drive one standard TTL load in the output mode. CA1 controls the latching of data on Port A input lines. CA1 is a high-impedance input only. CA2 represents one standard load in the input mode and will drive one standard TTL load in the output mode.

Port B data lines represent one standard TTL load in the input mode and will drive one standard TTL load in the output mode. Additionally, they are capable of sourcing 1.0 mA at 1.5V dc in the output mode. This feature permits the direct driving of Darlington transistor loads.

Figure 4.4 describes port register 0 (ORB/IRB). *Figure 4.5* describes Register 1 (ORA/IRA). *Figure 4.6* describes Registers 2 (DDRB) and 3 (DDRA). *Figure 4.7* shows the output circuits for Ports A and B. The input latching modes are selected by the auxiliary control register number 12 (see *Figure 4.8*).

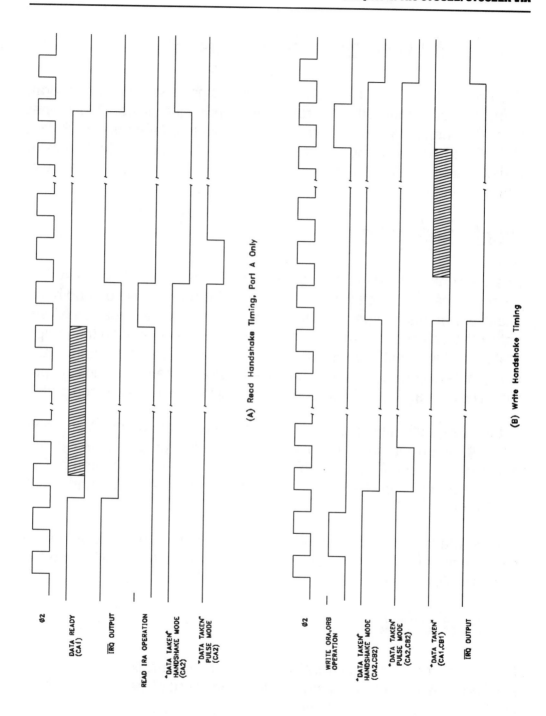

Figure 4.9. Handshake timing diagram for (A) Port A read only and (B) Ports A and B write.

Handshake Control of Data Transfers

Positive control of data transfers between the processor and peripherals is maintained through the operation of "handshake" lines. Port A lines CA1, CA2 handshake on both a read and a write operation. Port B lines CB1, CB2 handshake on a write operation only. *Figure 4.9A* illustrates Port A read-handshake timing, while *Figure 4.9B* illustrates write-handshake timing.

Read Handshake

Positive control of data transfers from peripheral devices into the system processor can be effectively accomplished through the use of read handshaking. For this, the peripheral device must provide a "data-ready" signal for the processor, signifying that valid data is available on the peripheral port. This signal interrupts the processor, which reads the data, causing a "data-taken" signal. The peripheral device then responds by providing new data. The process continues until the transfers are completed.

In the 6522, automatic read handshaking is available on the A port only. The CA1 interrupt input accepts the data-ready signal and CA2 generates the data-taken signal. The data-ready signal sets an internal flag, which may interrupt the processor or which may be polled under program control.

The data-taken signal can either be a pulse or a level, which is set low by the system processor and is cleared by the data-ready signal. These features are illustrated in Figure 4.9A.

Write Handshake

For write handshaking, the 6522 generates the data-ready signal and the peripheral device must respond with the data-taken signal. This is accomplished on both the A and B ports.

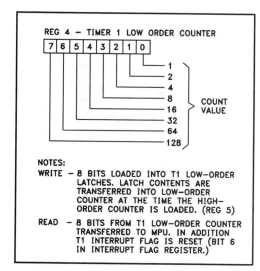

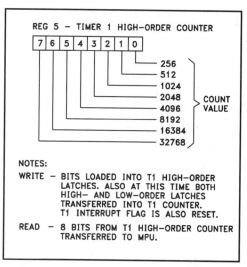

Figure 4.10. Register format for Timer 1 low-order counter.

Figure 4.11. Register format for Timer 1 high-order counter.

CA2 or CB2 will function as a data-ready output in either the handshake or pulse mode. CA1 or CB1 accept the data-taken signal from the peripheral device, setting the interrupt flag and clearing the data-ready output. This sequence is shown in Figure 4.9B.

Selection of the operating modes for CA1, CA2, CB1, and CB2 is accomplished by the peripheral control register, number 12, which is shown in Figure 4.8.

Latch and Counter Registers

We find the six registers that follow to be dedicated to latch and counter functions. From the block diagram of Figure 4.1 we see that the 6522 provides two timer/counter functions.

These differ in that Timer/Counter 1 is provided with both low-order and high-order latching, while Timer/Counter 2 has low-order latching only. A good question to ask at this time, though, is what is being timed, or counted.

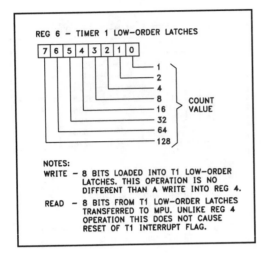

Figure 4.12. Register format for Timer 1 low-order latches.

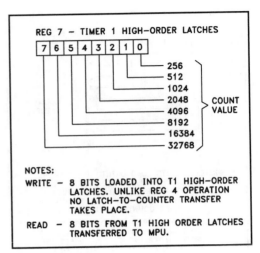

Figure 4.13. Register format for Timer 1 high-order latches.

Read and write decisions are always determined by the status of the R//W pin, pin number 22. So when we read "write" or "read" on the register drawings, we have to keep in mind the status of this input.

Timer/Counter 1 Functions

Interval Timer 1 consists of two eight-bit latches and a 16-bit counter. Register 4 contains the low-order bits; register 5 the high-order. These are shown in *Figures 4.10* and *4.11*, respectively. The latches hold the data to be loaded into the counter. Once loaded, the counter is decremented at the Ø2 clock rate. What this means is that, with each tick of the clock, the counter value is reduced by one count. When the value is reduced to zero, an interrupt flag will be set (Bit 6 of the interrupt flag register). Then /IRQ will go low, if the interrupt has been enabled.

If we enable the interrupt on loading the counter, we may time the interval between the enable and /IRQ going low. We are unable to observe this capability in this book.

Left to itself, the timer now disables any other interrupts, or will automatically transfer the latch content into the counter and continue with the decrementing. Additionally, we may program the timer to invert the output signal on a peripheral pin each time it passes through zero, or "times out." And now we have a source of square waves at some fraction of the Ø2 rate. This is set by the bit pattern loaded into the latches.

Since we're using the data bus for many differing operations, there has to be a means for selecting the desired function. We will get to this as we proceed.

Registers 6 and 7 are identified in *Figures 4.12* and *4.13* as Timer 1 low-order and high-order latches, respectively. The data entered into the latches is moved into the counter when given a load instruction. The latch data may also be read out to the microprocessor (MPU). In this book, that's us—*we* are a slow-speed, pretend MPU.

Timer 1 has two modes of operation of interest to us: the one-shot mode for interrupt generation and a free-running mode for square-wave generation. Let's take a look at these before moving on to Timer 2.

Timer 1 One-Shot Mode

This mode is provided for the generation of a single interrupt for each cycle of counter operation. The time interval between the "write T1C-H" operation (read load T1 high-order counter) and generation of the interrupt is a function of the data loaded into the counter. In addition to the interrupt, this timer can be programmed to yield a single negative-going pulse on Port B, pin PB7. With the output enabled (ACR7=1), a "write T1C-H" operation will cause PB7 to go low. When the timer times out, this pin will go high. The result is a pulse having a programmable width.

In this mode, writing to the high-order latch will have no effect on the timer operation. We do, however, want to ensure the low-order latch has the proper data prior to initiating the countdown with the write order. When data is written into the high-order counter, the T1 interrupt flag is cleared,

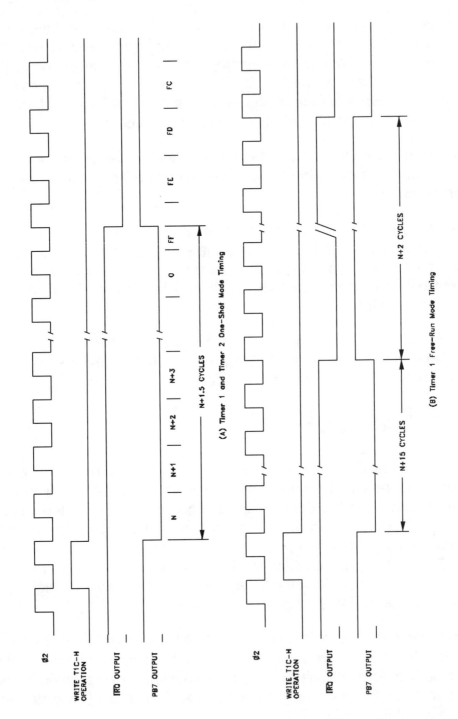

Figure 4.14. Timer timing diagrams: (A) Timer 1 and 2 one-shot mode; (B) Timer 1 free-running mode.

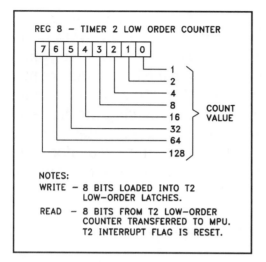

Figure 4.15. Register format for Timer 2 low-order counter.

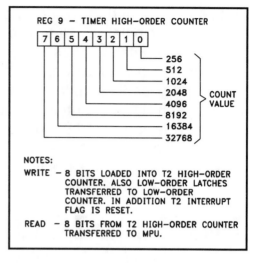

Figure 4.16. Register format for Timer 2 high-order counter.

the content of the low-order latch is transferred into the low-order counter, and counter decrementing is begun at the system (Ø2) clock rate.

If the PB7 output is enabled, then its output will go low on the clock cycle following the write instruction. When the counter value drops to zero, the T1 interrupt flag will be set, the /IRQ pin will go low, and the signal on PB7 will go high. The counter will continue to decrement at the clock rate. The T1 interrupt flag must be cleared before it can be set again. *Figure 4.14A* illustrates the system timing for the one-shot modes of Timers 1 and 2.

Timer 1 Free-Run Mode

This mode produces a continuous series of evenly spaced interrupts, resulting in a square wave appearing on PB7. In this mode, the interrupt flag is set and the signal on PB7 inverted each time the counter reaches zero. Now the counter does not continue to decrement as in the one-shot mode, but is set to the value stored in the latches instead, then continues to decrement from there. This is accomplished automatically; there is no need to rewrite the timer to maintain the cycle.

Both Timers 1 and 2 are re-triggerable in that rewriting the counter will always reinitialize the time-out interval. Should the rewrite occur before the counter reaches zero, the time-out can be prevented. Timer 1 will operate in this fashion, if data is written into the high-order counter (T1C-H). By loading the latches alone, the MPU can access the timer during each down-counting operation without affecting the time-out process. Instead, the data loaded into the latches will determine the length of the next time-out duration. With this feature, waveforms of varying complexity are easily generated. Timing for the free-running mode is shown in *Figure 4.14B*.

Timer/Counter 2 Functions

The remaining two latch/counter registers are numbers 8 and 9, shown in *Figures 4.15* and *4.16*, respectively. Figure 4.15 illustrates the Timer 2 low-order counter; Figure 4.16 the high-order counter.

Timer 2 functions as an interval counter in the one-shot mode only. It also functions as a counter for negative pulses on the Port PB6 pin. A single control bit is available in the auxiliary control register to select between these two modes.

This timer is comprised of a "write-only" low-order latch (T2L-L), a "read-only" low-order counter, and a read/write high-order counter. The counter registers perform as a 16-bit counter that decrements at a Ø2 rate.

Timer 2 One-Shot Mode

Timer 2 acts as an interval timer in the one-shot mode similar to Timer 1. Here the timer provides a single interrupt for each "write T2C-H" operation. After timing out, the counter will continue to decrement. But the setting of the interrupt flag will be disabled after the initial time-out, so it will not be set by the counter continuing to decrement through zero. We must rewrite T2C-H to enable setting the flag. The interrupt flag is cleared by reading T2C-L or by writing T2C-H. Timing for this timer's operation is also shown in Figure 4.14A.

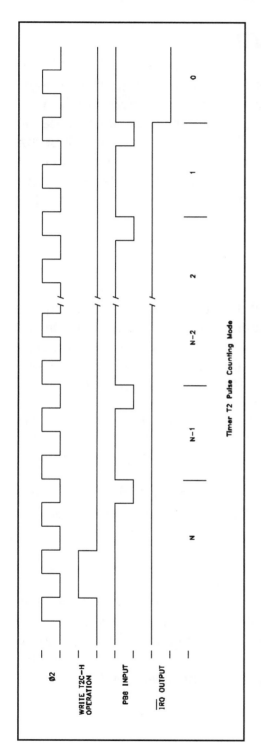

Timer T2 Pulse Counting Mode

Ø2

WRITE T2C–H
OPERATION

PB6 INPUT

IRQ OUTPUT

Figure 4.17. *(At left)* Timing diagram for Timer 2 pulse-counting mode.

REG 10 – SHIFT REGISTER

| 7 | 6 | 5 | 4 | 3 | 2 | 1 | 0 |

SHIFT
REGISTER
BITS

NOTES:
1. WHEN SHIFTING OUT BIT 7 IS THE FIRST
   BIT OUT AND SIMULTANEOUSY IS ROTATED
   BACK INTO BIT 0.
2. WHEN SHIFTING IN BITS INITIALLY ENTER
   BIT 0 AND AND ARE SHIFTED TOWARDS
   BIT 7.

Figure 4.18. Register format for the shift register.

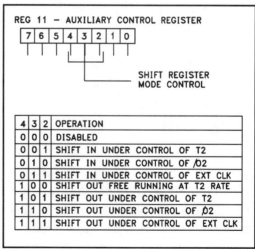

REG 11 – AUXILIARY CONTROL REGISTER

| 7 | 6 | 5 | 4 | 3 | 2 | 1 | 0 |

SHIFT REGISTER
MODE CONTROL

4	3	2	OPERATION
0	0	0	DISABLED
0	0	1	SHIFT IN UNDER CONTROL OF T2
0	1	0	SHIFT IN UNDER CONTROL OF Ø2
0	1	1	SHIFT IN UNDER CONTROL OF EXT CLK
1	0	0	SHIFT OUT FREE RUNNING AT T2 RATE
1	0	1	SHIFT OUT UNDER CONTROL OF T2
1	1	0	SHIFT OUT UNDER CONTROL OF Ø2
1	1	1	SHIFT OUT UNDER CONTROL OF EXT CLK

Figure 4.19. Register format for the auxiliary control register.

Timer 2 Pulse-Counting Mode

In this mode, Timer 2 serves to count a predetermined number of negative-going pulses on PB6. To do this, we first write a number into T2. Writing into T2C-H clears the interrupt flag and permits the counter to decrement each time a pulse is applied to PB6. The interrupt flag will be set when T2 reaches zero. The counter will continue to decrement with each pulse on PB6. But it is necessary to rewrite T2C-H to enable the interrupt flag to be set on each subsequent down count sequence. Timing for this mode is shown in *Figure 4.17*.

Shift Register Operation

The shift register (SR) performs serial data transfers into and out of the CB2 pin under control of an internal modulo-8 counter. Shift pulses can be applied to the CB1 pin from an external source—or when this mode is selected—from pulses generated internally within the 6522. The shift register bit pattern is provided by Register 10, shown in *Figure 4.18*.

Shift-register control is located in Register 11, the auxiliary control register (ACR), shown in *Figure 4.19*. The eight control modes that follow are defined in the ACR.

The control timing for shift-in and shift-out modes is depicted in *Figures 4.20* and *4.21*. Not included is Mode 000, for which the SR is disabled.

Mode 001

Mode 001, shown in *Figure 4.20A*, features shift-in under control of Timer T2. In this mode, the shifting rate is controlled by the low-order eight bits of T2. Shift pulses are generated on the CB1 pin to control shifting external devices. The time between transitions of this output clock is a function of the system clock (Ø2), period, and the contents of the low-order T2 latch.

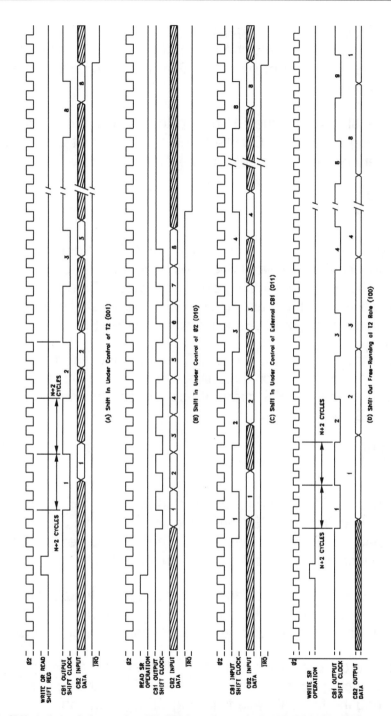

Figure 4.20. Shift register timing modes: (A) shift-in under control of T2 (001); (B) shift-in under control of f2 (010); (C) shift-in under control of external CB1 clock (011); (D) shift-out free-running at T2 rate (100).

The shifting operation is triggered by reading or writing the shift register. Data is shifted first into the low-order bit of SR and is then shifted into the next higher-order bit of the SR on the negative-going edge of each clock pulse.

The data is shifted into the SR during the Ø2 clock cycle following the positive-going edge of the CB1 clock pulse. After eight CB1 clock pulses, the SR interrupt flag will be set and the IRQ will go low.

Mode 010

Mode 010, shift-in under control of the clock, Ø2, is a direct function of the clock frequency (see *Figure 4.20B*). CB1 becomes an output, which generates shift pulses for controlling external devices. Timer 2 operates as an independent interval timer and has no effect on the shift register. The shifting operation is triggered by reading or writing the SR.

Data is shifted first into Bit 0 of the SR and is then shifted into the next higher-order bit of the SR on the trailing edge of each Ø2 clock pulse. After eight clock pulses, the SR interrupt flag will be set and the output clock pulses on CB1 will stop.

Mode 011

Mode 011, shown in *Figure 4.20C*, is shift-in under control of an external CB1 clock. In this mode, CB1 becomes an input. This allows an external input to load the SR at its own pace. The SR counter will interrupt the processor each time eight bits have been shifted in.

Yet the shift register counter does not stop the shifting operation; it acts simply as a pulse counter. Reading or writing the SR resets the interrupt flag and initializes the SR counter to count another eight pulses.

Note that the data is shifted during the first system clock cycle following the positive-going edge of CB1 shift pulse. Thus, the data must be held stable during the first full cycle following CB1 going high.

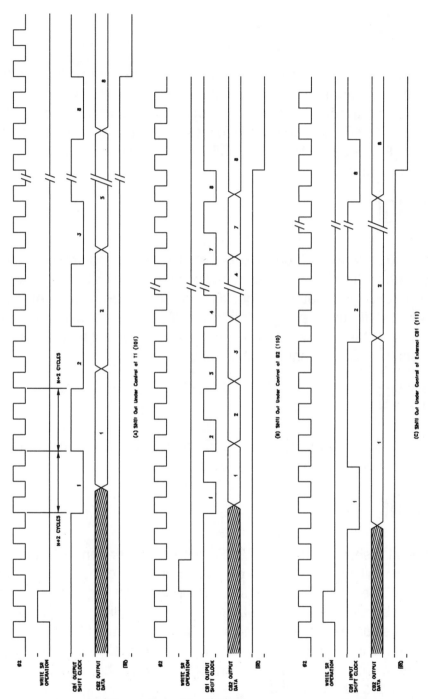

Figure 4.21. Shift register timing modes: (A) shift-out under control of T1 (101); (B) shift-out under control of f2 (110); (C) shift-out under control of external CB1 clock (111).

Mode 100

Shift-out free-running at T2 rate, Mode 100 (shown in *Figure 4.20D*) is very similar to Mode 101 in which the shifting rate is set by T2. In Mode 100, however, the SR counter does not stop the shifting operation. Since the SR Bit 7 (SR7) is recirculated back into Bit 0, the eight bits loaded into the SR will be clocked onto CB2 repetitively. In this mode, the SR clock is disabled.

Mode 101

In Mode 101 (shown in *Figure 4.21A*), shift-out under control of T2, the shift rate is controlled by T2. With each read or write of the shift register, however, the SR counter is reset and eight bits are shifted onto CB2. At the same time, eight shift pulses are generated on CB1 to control shifting in external devices. After the eight shift pulses, the shifting is disabled, the SR interrupt flag is set, and CB2 remains at the last data level.

Mode 110

In Mode 110 (shown in *Figure 4.21B*), shift-out under control of Ø2, the shift rate is controlled by the Ø2 clock.

Mode 111

In Mode 111 (shown in *Figure 4.21C*), shift-out under control of the external CB1 clock, shifting is controlled by pulses applied to the CB1 pin from an external source. The SR counter sets the SR interrupt flag each time it counts eight pulses, but it does not disable the shifting function. Each time the processor writes or reads the shift register, the SR interrupt flag is reset and the SR counter is initialized to begin counting the next eight shift pulses on pin CB1. After eight shift pulses, the interrupt flag is set. The register can then be loaded with the next byte of data.

Interrupt Operation

Three principal operations are involved with the interrupt: flagging the interrupt, enabling the interrupt, and signaling that an active interrupt exists within the 6522.

Interrupt flags are set for interrupting conditions that exist within the chip or from external inputs. Once set, these flags normally remain so until reset on servicing the interrupt. In practice, the MPU must determine the source of the interrupt by examining the flags from highest to lowest priority. This is done by reading the flag register into the MPU's accumulator, shifting the register data either right or left and using conditional branch instructions to detect the active interrupt. In this book, we will perform this procedure somewhat differently. We will read the flag register, but the shifting operation will not be performed.

Associated with each interrupt flag is an interrupt enable bit. This can be set or cleared by the MPU as appropriate to enable interrupting the processor from the corresponding flag.

If an interrupt flag is set to a logic 1 by the interrupting condition, and if the corresponding interrupt enable bit it set to a logic 1, then the interrupt request output (/IRQ) will go low. /IRQ is an open collector output that can be "write-or'd" with other devices within a system to interrupt the MPU.

In the 6522, the interrupt flags are contained in Register 13, the interrupt flag register (IFR), shown in *Figure 4.22*. In addition, Bit 7 of this register will appear as a logic 1 when an interrupt exists within the chip. In a system of several devices, this allows convenient polling to locate the source of the interrupt.

The IFR may be read directly by the processor. Individual flag bits may be cleared by writing a "1" into the appropriate bit of the IFR. When the correct chip select and register signals are input to the chip, the register contents are placed on the data bus. Bit 7 indicates the status of the IRQ output.

This bit corresponds to the logic function:

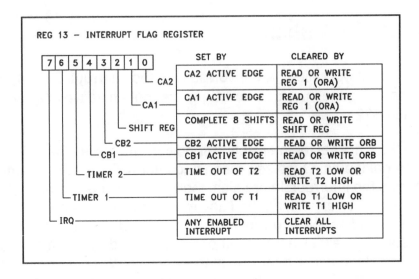

Figure 4.22. Register format for the interrupt flag register (IFR).

IRQ = IFR6xIER6 + IFR5xIER5 + IFR4xIER4 + IFR3xIER3 + IFR2xIER2 + IFR1xIER1 + IFR0xIER0

where
x = logical AND
+ = logical OR

The IFR Bit 7 is not a flag and is not directly cleared by writing a logic 1 into it. It can be cleared only by clearing all the register flags or disabling all the active interrupts.

For each interrupt flag in the IFR there is a corresponding bit in Register 14, the interrupt enable register (IER), shown in *Figure 4.23*. Selected bits in this register can be set or cleared to facilitate controlling individual interrupts without affecting others. This is accomplished by writing to IER address 1110. If Bit 7 of the data placed on the data bus during this write operation is a 0, each 1 in Bits 6 through 0 clears the corresponding bit in the IER. For each 0 in these bits the corresponding bit is not affected.

Setting selected bits in the IER is accomplished by writing to the same address with Bit 7 in the data word set to logic 1. In this case, each 1 in

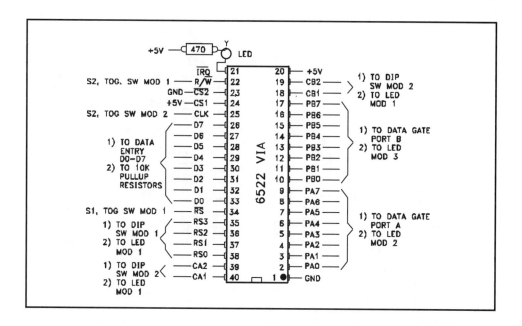

Figure 4.24 The wiring plan for the author's bench-top module layout for operation of the 6522 VIA.

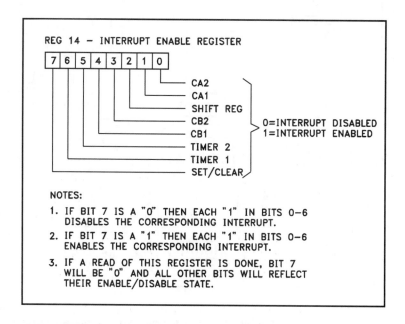

Figure 4.23. Register format for the interrupt enable register (IER).

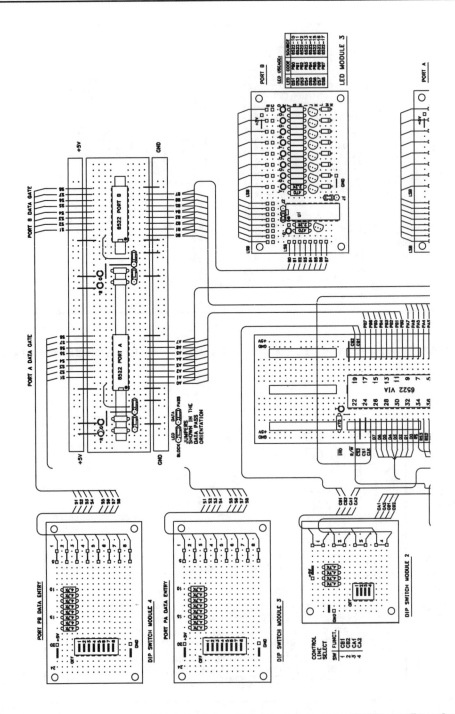

Figure 4.25. *(Continued on next page.)* **The author's bench- top module layout and wiring configuration for operation of the 6522 VIA.**

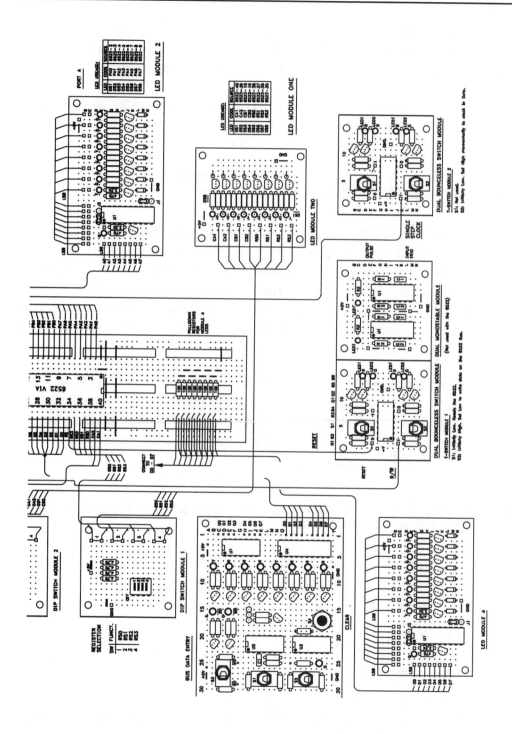

Bits 6 through 0 will set the corresponding bit. For each 0 the corresponding bit will be unaffected. This is a very convenient method of maintaining individual interrupt control during system operations.

In addition to setting and clearing IER bits, the processor can read the contents of this register by placing the proper address on the register select and chip select inputs with the R//W line high.

Bench-Top Operation of the 6522

Figures 4.24 and *4.25* illustrate my own layout for bench-top operation of the 6522. Figure 4.24 is a planning drawing provided as an aid in our wiring of the test configuration shown in Figure 4.25. Here is shown the overall arrangement and the interconnecting wiring linking the modules and the 6522. Since your modules may be constructed differently from mine, the appearance of your setup may not be the same, but the functions must be provided.

I strongly recommend fastening your modules to a supporting base. I have found corrugated cardboard to be a cost-effective and most satisfactory material for this. Just slice up a cardboard box having sides of an adequate size. Holes for attaching screws can be easily punched and the modules secured in place.

Table 4.4. Instructions on usage of the data entry and R//W toggle switches found on the bus data-entry module and dual bounceless switch module.

```
HOW TO USE THE DATA ENTRY AND 6522 R/W TOGGLES
1. With the clock switch set LOW and both R/W switches in the
   Read position, set the data pattern on the Bus Data Entry Module.
   Observe that the pattern is correct on the Module output LEDs.
2. Set the Register Selection DIP switches for the required register.
3. To write to the 6522 Bus:
   Set both R/W switches to write.
   Note that the write pattern is present on LED Module 4.
   Do a single clock step.
4. Return both R/W switches to the Read position.
5. To verify that the write was successfully read do a single clock step.
   The data pattern written into the bus appears on LED Module 4
   when the clock step is HIGH. The LEDs are blank with the clock LOW.
```

I use number 22 solid insulated wire of various colors and lengths (a practice of many years), so my stock is well organized by lengths. Kits of preformed short lengths are available, but you will need longer lengths as well. Use of more than a single color is helpful when tracing the wiring later. Do not be either too stingy or too generous with wire lengths. You want to be able to route the wiring around the switches and display LEDs without overdoing it. As you can see, there is a substantial amount of wiring to be done. Take care with the connections; it is all too easy to plug into an adjacent point. Check yourself as you proceed. It's easier to catch mistakes now than later on when it will be very mysterious.

Review the switch usage procedures of *Table 4.4* carefully before beginning any operations. In particular, have an understanding of read and write as applied to the 6522. If you have constructed the data-entry module as described in Chapter 1, then most likely you have operated it sufficiently at this time to need no further instruction on its use. If not, now is a good time for a bit of practice. If you are using a switching arrangement of your own devising, that is well and good—provided it meets the needs of entering data on the bus for write operations and isolating the bus when in the read mode. This requirement applies to the two peripheral ports as well. The port gating circuitry for isolating the DIP switches is provided in Figure 1.11 of Chapter 1.

Figures 4.26 through *4.34* describe procedures for operating the 6522 using the bench-top setup of *Figure 4.25*. These nine procedures do not cover all the possible modes of the device operation. They will, however, provide insight into the register and timer functioning such that you can devise additional procedures of your own should this device be of greater interest to you.

The first procedure, shown in Figure 4.26, provides for inputting data to peripheral Port A with CA1 and CA2 handshaking. Each procedure is designed to be self-explanatory. Even so, it would be helpful to review the text on related features and the procedures with care before and/or during your operation of the device. Review Table 4.4 for an understanding of how to use the read and write toggle switches. Also note the need to pass or block data from the PA and PB DIP switch modules as required with each procedure.

TO INPUT DATA ON PORT A:

CAUTION: Review Table 4.___ before beginning this procedure to ensure correct R/W switch and DIP switch useage.

1. Set Peripheral Control Register 12 for CA1, CA2 as shown. Ensure CA1 and CA2 both set to 1.

   RS3-RS0 = 1100 REG 12 SETUP ☐

   OFF
   RS0 ▢ 0
   RS1 ▢ 0
   RS2 ▢ 1
   RS3 ▢ 1   ☐
   1 0
   REGISTER 12
   DIP SW SETTINGS

7 6 5 4 3 2 1 0	OPERATION
0 0 0 0 0 0 1 0	CA1 NEG ACTIVE EDGE CA2 INPUT NEG ACTIVE EDGE

   BUS DATA ENTRY = 0000 0010 ☐
   SINGLE STEP CLOCK INPUT ☐

   Jumper the A Port Data Gate for source pass through.
   ☐

   PORT A DIP SWITCH SETTING

   OFF
   ▢ +V
   ▢ +V
   ▢ GND
   ▢ +V
   ▢ +V
   ▢ GND
   ▢ +V
   ▢ +V   ☐

   OFF
   CB1 ▢ 1
   CB2 ▢ 1
   CA1 ▢ 1
   CA2 ▢ 1   ☐
   1 0
   CA, CB
   SELECT
   DIP SW

   PORT A DATA
   ENTRY DIP SWITCH

   Note: It is instructive to change the data switch settings after each single step and observe the new data present on the Bus as seen on the LED Module 4 display.

2. Select Data Direction Register 3 for Port A Data Read

   RS3-RS0 = 0011 REG 3 SETUP ☐

   OFF
   RS0 ▢ 1
   RS1 ▢ 1
   RS2 ▢ 0
   RS3 ▢ 0   ☐
   1 0
   REGISTER 3
   DIP SW SETTINGS

7 6 5 4 3 2 1 0	OPERATION
0 0 0 0 0 0 0 0	Set Port A for Data Input

   BUS DATA ENTRY = 0000 0000 ☐
   SINGLE STEP CLOCK INPUT ☐

3. Select Interrupt Enable Register 14 to enable control CA2.

   RS3-RS0 = 1110 REG 14 SETUP ☐

   OFF
   RS0 ▢ 0
   RS1 ▢ 1
   RS2 ▢ 1
   RS3 ▢ 1   ☐
   1 0
   REGISTER 14
   DIP SW SETTINGS

7 6 5 4 3 2 1 0	OPERATION
1 0 0 0 0 0 0 1	Enable Interrupt CA2

   BUS DATA ENTRY = 1000 0001 ☐
   SINGLE STEP CLOCK INPUT ☐

4. Select Input Register IRA for Data Entry. Read Port A.

   RS3-RS0 = 0001 REG 1 SETUP ☐

   OFF
   RS0 ▢ 1
   RS1 ▢ 0
   RS2 ▢ 0
   RS3 ▢ 0   ☐
   1 0
   REGISTER 1
   DIP SW SETTINGS

7 6 5 4 3 2 1 0	OPERATION
	Input Data, port A

   BUS DATA ENTRY = (DIP SW) ☐
   CAUTION: Both R/W switches must be in R position.
   SINGLE STEP CLOCK INPUT ☐

   Note that data is displayed on LED Module 4 only while the clock is HIGH. DO NOT RESET THE CLOCK!

   NOTE:

   The two R/W switches are to remain in the R position for steps 4 through 7.

   These steps may be repeated using new Port A DIP switch settings.

5. Set CA1, CA2 to 1.
   With Clock still HIGH, Port A Data displayed, Reset CA1, IRQ goes LOW.
   Set the Clock Low
   IRQ remains LOW.

6. Select Interrupt Flag Register 13 to observe CA2 Flag.

   RS3-RS0 = 1101 REG 13 SETUP ☐

   OFF
   RS0 ▢ 1
   RS1 ▢ 0
   RS2 ▢ 1
   RS3 ▢ 1   ☐
   1 0
   REGISTER 13
   DIP SW SETTINGS

7 6 5 4 3 2 1 0	OPERATION
1 0 0 0 0 0 0 1	

   OBSERVE FLAG OPERATION ☐
   SINGLE STEP CLOCK INPUT ☐

7. Read Register 13
   Bits 7, 0 = 1
   Do a second read
   Bits 7, 0 = 0
   IRQ goes HIGH on Clock LOW

Figure 4.26.   Procedure for inputting data to peripheral Port A with CA1, CA2 handshaking.

```
TO INPUT DATA ON PORT B:                                    Jumper the B Port Data
                                                            Gate for source pass through.
Note that there is no handshaking on a READ                 □
from the B Peripheral Port.
The procedure below may be carried out as part
of the previous routine for reading Port A if desired.
                                                            PORT B DIP SWITCH SETTING

1. Select Data Direction Register 2 for Port B              OFF
   RS3-RS0 = 0010 REG 2 SETUP □                                     1
   OFF        ┌─────────────────────────────────────┐              1
   RS0 [ ▭ ] 0 │ 7 6 5 4 3 2 1 0 │ OPERATION        │              1
   RS1 [ ▭ ] 1 │ 0 0 0 0 0 0 0 0 │ Set Port B for Data Input │     0
   RS2 [ ▭ ] 0 └─────────────────────────────────────┘             1
   RS3 [ ▭ ] 0  □ BUS DATA ENTRY = 0000 0000 □                     0
        1  0       SINGLE STEP CLOCK INPUT   □                     1
     REGISTER 2                                                    1
     DIP SW SETTINGS                                           1  0
                                                            PORT B DATA
2. Select Input Register IRB for Data Entry.                ENTRY DIP SWITCH

   Set the DATA ENTRY R/W Toggle to R.  □                   Note: It is instructive to change the
   RS3-RS0 = 0000 REG 0 SETUP □                                   data switch settings after each
   OFF        ┌─────────────────────────────────────┐            single step and observe the
   RS0 [ ▭ ] 0 │ 7 6 5 4 3 2 1 0 │ OPERATION        │            change on the 6522 data bus.
   RS1 [ ▭ ] 0 │                 │ Data input, port B│
   RS2 [ ▭ ] 0 └─────────────────────────────────────┘
   RS3 [ ▭ ] 0  □ BUS DATA ENTRY (DIP SW)    □
        1  0       SINGLE STEP CLOCK INPUT   □
     REGISTER 0    CAUTION: Both R/W switches must be in R position.
     DIP SW SETTINGS Note that data is displayed on LED Module 4 only
                    when the clock is HIGH.
```

Figure 4.27. Procedure for inputting data to peripheral Port B. Note that there is no handshaking on a read from this port.

The next procedure, shown in *Figure 4.27*, is for inputting data to peripheral Port B. This is a much simpler procedure in that there is no read handshaking with this port.

With the third procedure, shown in *Figure 4.28*, data is outputted from peripheral Port A. In this mode, there is Port A handshaking with CA1 and CA2.

The fourth procedure, shown in *Figure 4.29*, is similar to the third in that data is outputted from peripheral port. In this mode, there is Port B handshaking with CB1 and CB2.

In the fifth procedure, shown in *Figure 4.30*, we set up Timer 1 for operation in the square wave mode. With this mode, when operating with software,

TO OUTPUT DATA ON PORT A:

CAUTION: Review Table 4.__ before beginning this procedure
to ensure correct R/W switch and DIP switch useage.

Jumper the A Port Data Gate
to block the DIP switch input. ☐

1. Set Peripheral Control Register 12 for CA1, CA2 as shown.
   Ensure CA1 and CA2 both set to 1. ☐
   RS3-RS0 = 1100 REG 12 SETUP ☐

OFF		7 6 5 4 3 2 1 0	OPERATION
RS0	0	0 0 0 0 0 0 1 0	Independent Interrupt
RS1	0		Input – Negative Edge
RS2	1		
RS3	1 ☐		

1 0
REGISTER 12
DIP SW SETTINGS

BUS DATA ENTRY = 0000 0010 ☐
SINGLE STEP CLOCK INPUT ☐

OFF	
CB1	1
CB2	1
CA1	1
CA2	1 ☐

1 0
CA, CB
SELECT
DIP SW

2. Select Data Direction Register 3 for Port A Data Write.
   RS3-RS0 = 0011 REG 3 SETUP ☐

OFF		7 6 5 4 3 2 1 0	OPERATION
RS0	1	1 1 1 1 1 1 1 1	Set Port A for Data Output
RS1	1		
RS2	0		
RS3	0 ☐		

1 0
REGISTER 3
DIP SW SETTINGS

BUS DATA ENTRY = 1111 1111 ☐
SINGLE STEP CLOCK INPUT ☐

3. Select Interrupt Enable Register 14 to enable control CA2.
   RS3-RS0 = 1110 REG 14 SETUP ☐

OFF		7 6 5 4 3 2 1 0	OPERATION
RS0	0	1 0 0 0 0 0 0 1	Enable Interrupt CA2
RS1	1		
RS2	1		
RS3	1 ☐		

1 0
REGISTER 14
DIP SW SETTINGS

BUS DATA ENTRY = 1000 0001 ☐
SINGLE STEP CLOCK INPUT ☐

4. Select Output Register ORA for Data Output. Write to Port A.
   RS3-RS0 = 0001 REG 1 SETUP ☐

OFF		7 6 5 4 3 2 1 0	OPERATION
RS0	1		Output Data to Port A
RS1	0		
RS2	0		
RS3	0 ☐		

1 0
REGISTER 1
DIP SW SETTINGS

Place Data on Bus using Data Entry Module ☐
Note that both R/W switches must be in W.
SINGLE STEP CLOCK INPUT ☐
DO NOT RESET THE CLOCK!

NOTE:
The two R/W switches are
to remain in the W position
for steps 4 through 7.
These steps may be repeated
with new data entered on the
bus.

5. Set CA1, CA2 to 1.
   With Clock still HIGH, Port A
   Data displayed. Reset CA1.
   IRQ goes LOW.
   Set the Clock Low
   IRQ remains LOW.
7. Read Register 13
   Bits 7, 0 = 1
   Do a second read
   Bits 7, 0 = 0
   IRQ goes HIGH on Clock LOW.

6. Select Interrupt Flag Register 13 to observe CA2 Flag.
   RS3-RS0 = 1101 REG 13 SETUP ☐

OFF		7 6 5 4 3 2 1 0	OPERATION
RS0	1	1 0 0 0 0 0 0 1	
RS1	0		
RS2	1		
RS3	1 ☐		

1 0
REGISTER 13
DIP SW SETTINGS

OBSERVE FLAG OPERATION ☐
SINGLE STEP CLOCK INPUT ☐

Figure 4.28. Procedure for outputting data from peripheral Port A with CA1, CA2 handshaking.

TO OUTPUT DATA AT PORT B:

CAUTION: Review Table 4.__ before beginning this procedure to ensure correct R/W switch and DIP switch useage.

Jumper the B Port Data Gate to block the DIP switch input.

1. Set Peripheral Control Register 12 for CB1, CB2 as shown.

   Ensure CB1 and CB2 both set to 1. ☐

   RS3-RS0 = 1100 REG 12 SETUP ☐

   OFF
   RSO [ ] 0
   RS1 [ ] 0
   RS2 [ ] 1
   RS3 [ ] 1
        1 0
   REGISTER 12
   DIP SW SETTINGS

7 6 5 4 3 2 1 0	OPERATION
0 0 1 0 0 0 0 0	Independent Interrupt Input – Negative Edge

   BUS DATA ENTRY = 0000 0010 ☐
   SINGLE STEP CLOCK INPUT ☐

   OFF
   CB1 [ ] 1
   CB2 [ ] 1
   CA1 [ ] 1
   CA2 [ ] 1 ☐
        1 0
   CA, CB
   SELECT
   DIP SW

2. Select Data Direction Register 2 for Port B Data Write.
   RS3-RS0 = 0010 REG 12 SETUP ☐

   OFF
   RSO [ ] 0
   RS1 [ ] 1
   RS2 [ ] 0
   RS3 [ ] 0
        1 0
   REGISTER 2
   DIP SW SETTINGS

7 6 5 4 3 2 1 0	OPERATION
1 1 1 1 1 1 1 1	Set Port B for Data Output

   BUS DATA ENTRY = 1111 1111 ☐
   SINGLE STEP CLOCK INPUT ☐

3. Select Interrupt Enable Register 14 to enable control CB2.
   RS3-RS0 = 1110 REG 14 SETUP ☐

   OFF
   RSO [ ] 0
   RS1 [ ] 1
   RS2 [ ] 1
   RS3 [ ] 1
        1 0
   REGISTER 14
   DIP SW SETTINGS

7 6 5 4 3 2 1 0	OPERATION
1 0 0 0 1 0 0 0	Enable Interrupt CB2

   BUS DATA ENTRY = 1000 1000 ☐
   SINGLE STEP CLOCK INPUT ☐

4. Select Output Register ORB for Data Entry. Write to Port B.
   RS3-RS0 = 0000 REG 0 SETUP ☐

   OFF
   RSO [ ] 0
   RS1 [ ] 0
   RS2 [ ] 0
   RS3 [ ] 0
        1 0
   REGISTER 0
   DIP SW SETTINGS

7 6 5 4 3 2 1 0	OPERATION
	Output Data to Port B

   Place Data on Bus using Data Entry Module ☐
   Note that both R/W switches must be in W.
   SINGLE STEP CLOCK INPUT ☐
   DO NOT RESET THE CLOCK!

   NOTE:

   The two R/W switches are to remain in the W position for steps 4 through 7.

   These steps may be repeated using new data entered on the bus.

5. Set CB1, CB2 to 1.
   With Clock still HIGH, Port B Data displayed, Reset CB1.
   IRQ goes LOW.
   Set the Clock Low
   IRQ remains LOW.

6. Select Interrupt Flag Register 13 to observe CB2 Flag.
   RS3-RS0 = 1101 REG 13 SETUP ☐

   OFF
   RSO [ ] 1
   RS1 [ ] 0
   RS2 [ ] 1
   RS3 [ ] 1
        1 0
   REGISTER 13
   DIP SW SETTINGS

7 6 5 4 3 2 1 0	OPERATION
1 0 0 0 0 0 0 1	

   OBSERVE FLAG OPERATION ☐
   SINGLE STEP CLOCK INPUT ☐

7. Read Register 13
   Bits 7, 0 = 1
   Do a second read
   Bits 7, 0 = 0
   IRQ goes HIGH on Clock LOW.

Figure 4.29. Procedure for outputting data from peripheral Port B with CB1, CB2 handshaking.

TIMER 1 SETUP AND OPERATION FOR SQUARE WAVE MODE

1. Block the Port B Data Gate (red LED on).
Reset the 6522 (automatic if powering up).
Note: This clears the registers and places PB7 in the High state.

2. Select the Auxiliary Control Register 11. Enable Timer 1, Port PB (bits 7, 6).

RS3-RS0 = 1011 REG 11 SETUP

REGISTER 11 DIP SW SETTINGS

7	6	5	4	3	2	1	0	OPERATION
1	1	0	0	0	0	0	0	Generate continous interrupts a square wave output on PB7.

BUS DATA ENTRY = 1100 0000
SINGLE STEP CLOCK INPUT

3. Select the Interrupt Enable Register 14. Enable Time 1 Interrupt (bits 7, 6).

RS3-RS0 = 1110 REG 14 SETUP

REGISTER 14 DIP SW SETTINGS

7	6	5	4	3	2	1	0	OPERATION
1	1	0	0	0	0	0	0	

BUS DATA ENTRY = 1100 0000
SINGLE STEP CLOCK INPUT

4. Write the desired time into Register 4, Timer 1 Low-Order Counter.
Note: Since we are single stepping it is in our interest to keep this value low. Two clock steps are suggested, as shown.

RS3-RS0 = 0100 REG 4 SETUP

REGISTER 4 DIP SW SETTINGS

7	6	5	4	3	2	1	0	OPERATION
0	0	0	0	0	0	1	0	Two clock steps

BUS DATA ENTRY = 0000 0010
SINGLE STEP CLOCK INPUT

Note: We are actually writing into the low-order latches. This value will transfer to the low order counter when we write into the high order, at T1C-H, as the next step.

5. Select the T1 High-Order Counter Register 5.

RS3-RS0 = 0101 REG 5 SETUP

REGISTER 5 DIP SW SETTINGS

7	6	5	4	3	2	1	0	OPERATION
0	0	0	0	0	0	0	0	Write into high-order counter. Transfer low-order latch into low-order counter. Reset the T1 Interrupt Flag.

BUS DATA ENTRY = 0000 0000
SINGLE STEP CLOCK INPUT
PB7 GOES LOW

Note: If we write any value other than zero it will have to be counted down before observing the operation of PB7 and IRQ.

6. Select Interrupt Flag Register 15 to observe Timer 1. READ ONLY. Do not write to this register.

RS3-RS0 = 1101 REG 13 SETUP

REGISTER 13 DIP SW SETTINGS

7	6	5	4	3	2	1	0	OPERATION
1	1	0	0	0	0	0	0	Time out of T1

BUS DATA DISPLAY = 1100 0000
SINGLE STEP CLOCK INPUT

Continue the single step clocking.
On TIME OUT IRQ goes Low.
PB7 inverts.
Flag bits 7, 6 ON while Clock High
When Clock drops:
IRQ resets, Flag bits off

Continue the single step clocking while observing the cycle pattern of the Flag Register, IRQ, and PB7. Try revising the input to Register 4 and note the change in the cycle timing. In operation with a MPU this provides a means for complex pattern generation.

Figure 4.30. Procedure for setting up Timer T1 for operation in the square-wave mode.

TIMER 1 SETUP AND OPERATION FOR ONE-SHOT MODE

1. Block the Port B Data Gate (red LED on).
   Reset the 6522 (automatic if powering up).
   Note: This clears the registers and places PB7 in the High state.

2. Select the Auxillary Control Register 11. Enable Timer 1, Port PB (bit 7).

RS3-RS0 = 1011 REG 11 SETUP

		7 6 5 4 3 2 1 0	OPERATION
OFF			
RS0	1	1 0 0 0 0 0 0 0	Generate a single interrupt and output pulse on PB7 for each T1 load sequence.
RS1	1		
RS2	0		
RS3	1		

BUS DATA ENTRY = 1000 0000
SINGLE STEP CLOCK INPUT

1 0
REGISTER 11
DIP SW SETTINGS

3. Select the Interrupt Enable Register 14. Enable Time 1 Interrupt (bits 7, 6)

RS3-RS0 = 1110 REG 14 SETUP

		7 6 5 4 3 2 1 0	OPERATION
OFF			
RS0	0	1 1 0 0 0 0 0 0	
RS1	1		
RS2	1		
RS3	1		

BUS DATA ENTRY = 1100 0000
SINGLE STEP CLOCK INPUT

1 0
REGISTER 14
DIP SW SETTINGS

4. Write the desired time into Register 4, Timer 1 Low-Order Counter.
   Note: Since we are single stepping it is in our interest to keep this value low. Two clock steps are suggested, as shown.

RS3-RS0 = 0100 REG 4 SETUP

		7 6 5 4 3 2 1 0	OPERATION
OFF			
RS0	0	0 0 0 0 0 0 1 0	Two clock steps
RS1	0		
RS2	1		
RS3	0		

BUS DATA ENTRY = 0000 0010
SINGLE STEP CLOCK INPUT

1 0
REGISTER 4
DIP SW SETTINGS

Note: We are actually writing into the low-order latches. This value will transfer to the low order counter when we write into the high order. T1C-H, as the next step.

5. Select the T1 High-Order Counter Register 5.

RS3-RS0 = 0101 REG 5 SETUP

		7 6 5 4 3 2 1 0	OPERATION
OFF			
RS0	1	0 0 0 0 0 0 0 0	Write into high-order counter. Transfer low-order latch into low-order counter. Reset the T1 Interrupt Flag.
RS1	0		
RS2	1		
RS3	0		

BUS DATA ENTRY = 0000 0000
SINGLE STEP CLOCK INPUT
PB7 GOES LOW

1 0
REGISTER 5
DIP SW SETTINGS

Note: If we write any value other than zero it will have to be counted down before observing the operation of PB7 and IRQ.

6. Select the Interrupt Flag Register to observe Timer 1.
   READ ONLY. Do not write to this register.

RS3-RS0 = 1101 REG 13 SETUP

		7 6 5 4 3 2 1 0	OPERATION
OFF			
RS0	1	1 1 0 0 0 0 0 0	Time out of T1
RS1	0		
RS2	1		
RS3	1		

BUS DATA DISPLAY = 1100 0000
SINGLE STEP CLOCK INPUT

1 0
REGISTER 13
DIP SW SETTINGS

Continue the single step clocking.
On TIME OUT IRQ goes Low.
PB7 goes High.
Flag bits 7, 6 ON while Clock High
When Clock drops:
IRQ resets, Flag bits OFF

NOTE: To continue return to step 4 and repeat the procedure from that point. You may change the entries for T1C-L and T1C-H as desired for the new cycle.

Figure 4.31. Procedure for setting up Timer T1 for operation in the one-shot mode.

TIMER 2 SETUP AND OPERATION FOR ONE-SHOT MODE

1. Block the Port B Data Gate (red LED on).
Reset the 6522 (automatic if powering up).
Note: This to clear the registers.

2. Select the Auxiliary Control Register 11, Enable Timer 2.
RS3-RS0 = 1011 REG 11 SETUP

7	6	5	4	3	2	1	0	OPERATION
0	0	0	0	0	0	0	0	Write T2L-L Read T2C-L Clear the Interrupt Flag

BUS DATA ENTRY = 0000 0000
SINGLE STEP CLOCK INPUT

REGISTER 11
DIP SW SETTINGS

3. Select the Interrupt Enable Register 14, Enable Timer 2 Interrupt (bits 7, 5).
RS3-RS0 = 1110 REG 14 SETUP

7	6	5	4	3	2	1	0	OPERATION
1	0	1	0	0	0	0	0	Enable Timer 2 Interrupt

BUS DATA ENTRY = 1010 0000
SINGLE STEP CLOCK INPUT

REGISTER 14
DIP SW SETTINGS

4. Write the desired time into Register 8, Timer 1 Low-Order Counter.
Note: Since we are single stepping it is in our interest to keep this value low. Two clock steps are suggested, as shown.
RS3-RS0 = 1000 REG 8 SETUP

7	6	5	4	3	2	1	0	OPERATION
0	0	0	0	0	0	1	0	Two clock steps

BUS DATA ENTRY = 0000 0010
SINGLE STEP CLOCK INPUT

REGISTER 8
DIP SW SETTINGS

Note: We are actually writing into the low-order latches. This value will transfer to the low order counter when we write into the high order, T2C-H, as the next step.

5. Select the T2 High-Order Counter Register 9.
RS3-RS0 = 1001 REG 9 SETUP

7	6	5	4	3	2	1	0	OPERATION
0	0	0	0	0	0	0	0	Write into high-order counter. Transfer low-order latch into low-order counter. Reset the T2 Interrupt Flag.

BUS DATA ENTRY = 0000 0000
SINGLE STEP CLOCK INPUT

REGISTER 9
DIP SW SETTINGS

Note: If we write any value other than zero it will have to be counted down before observing the operation of IRQ.

6. Select Interrupt Flag Register 13 to observe Timer 2.
READ ONLY. Do not write to this register.
RS3-RS0 = 1101 REG 13 SETUP

7	6	5	4	3	2	1	0	OPERATION
1	0	1	0	0	0	0	0	Time out of T2

BUS DATA DISPLAY = 1010 0000
SINGLE STEP CLOCK INPUT

REGISTER 13
DIP SW SETTINGS

Continue the single step clocking.
On TIME OUT IRQ goes Low.
Flag bits 7, 5 ON while Clock High
When Clock drops; IRQ resets, Flag bits OFF.

NOTE: To continue return to step 4 and repeat the procedure from that point. You may change the entries for T2C-L and T2C-H as desired for the new cycle.

Figure 4.32. Procedure for setting up Timer T2 for operation in the one-shot mode.

SHIFT REGISTER PROCEDURE FOR SHIFT IN UNDER CONTROL OF THE/O2 CLOCK

Note: Refer to the timing diagram of Figure 4.19(b).
Block the Port B Data Gate (Red LED ON)

1. Select the Auxiliary Control Register 11
Mode 010 Shift in under Ø2 control
RS3-RS0 = 1011 REG 11 SETUP ☐

OFF
RSO ▭ 1
RS1 ▭ 1
RS2 ▭ 0
RS3 ▭ 1 ☐
   1 0
REGISTER 11
DIP SW SETTINGS

7 6 5 4 3 2 1 0	OPERATION
0 0 0 0 1 0 0 0	Shift in under the clock

BUS DATA ENTRY 0000 1000 ☐
SINGLE STEP CLOCK INPUT ☐

2. Select the Interrupt Enable Register 14

RS3-RS0 = 1110 REG 14 SETUP ☐

OFF
RSO ▭ 0
RS1 ▭ 1
RS2 ▭ 1
RS3 ▭ 1 ☐
   1 0
REGISTER 14
DIP SW SETTINGS

7 6 5 4 3 2 1 0	OPERATION
1 0 0 0 0 1 0 0	Shift Register Interrupt

BUS DATA ENTRY 1000 0100 ☐
SINGLE STEP CLOCK INPUT ☐

3. Select the Shift Register to trigger the Shift Operation.

RS3-RS0 = 1010 REG 10 SETUP ☐

OFF
RSO ▭ 0
RS1 ▭ 1
RS2 ▭ 0
RS3 ▭ 1 ☐
   1 0
REGISTER 10
DIP SW SETTINGS

7 6 5 4 3 2 1 0	OPERATION
No data entry	Read the register

STEP ONE CLOCK CYCLE ONLY ☐
CAUTION: Both R/W switches must be in R position.

4. Select the Interrupt Flag Register 13.

RS3-RS0 = 1101 REG 13 SETUP ☐

OFF
RSO ▭ 1
RS1 ▭ 0
RS2 ▭ 1
RS3 ▭ 1 ☐
   1 0
REGISTER 13
DIP SW SETTINGS

7 6 5 4 3 2 1 0	OPERATION
1 0 0 0 0 1 0 0	Complete 8 shifts

BUS DATA DISPLAY 1000 0100 ☐
Do NOT write to this register ☐

Following the single clock of step 3 the CB1 LED is ON.
For each toggle of the clock switch the CB1 alternates between ON and OFF.
When the CB1 LED is OFF, IFR bit 4 is ON with clock switch High.
On completion of the shift:
   IFR bits 2 and 7 ON ☐
   IRQ is Low ☐

CONTROL LINE DIP SWITCH SEQUENCE

STEP 1 ☐
OFF
CB1 ▭ Open
CB2 ▭ O
CA1 ▭ 1
CA2 ▭ 1
   1 0

STEP 5 ☐
OFF
CB1 ▭ Open
CB2 ▭ O
CA1 ▭ 1
CA2 ▭ 1
   1 0

STEP 2 ☐
OFF
CB1 ▭ Open
CB2 ▭ O
CA1 ▭ 1
CA2 ▭ 1
   1 0

STEP 6 ☐
OFF
CB1 ▭ Open
CB2 ▭ O
CA1 ▭ 1
CA2 ▭ 1
   1 0

STEP 3 ☐
OFF
CB1 ▭ Open
CB2 ▭ O
CA1 ▭ 1
CA2 ▭ 1
   1 0

STEP 7 ☐
OFF
CB1 ▭ Open
CB2 ▭ O
CA1 ▭ 1
CA2 ▭ 1
   1 0

STEP 4 ☐
OFF
CB1 ▭ Open
CB2 ▭ O
CA1 ▭ 1
CA2 ▭ 1
   1 0

STEP 8 ☐
OFF
CB1 ▭ Open
CB2 ▭ O
CA1 ▭ 1
CA2 ▭ 1
   1 0

Be sure that CB1 is disconnected from
its DIP switch.
The 6522 CB1 pin 18 connection to
LED Module 2 must be present.

O = 0 as shown.
Fill in the circle to indicate a one if
desired.
Suggest CB2 be kept at "1" for first test.
CA1, CA2 have no role to play.

5. Repeat steps 3 and 4 for ☐
another cycle with new
CB2 switch settings.

Figure 4.33. Procedure for setting up the shift register for shift-in under control of the f2 clock.

we have the ability to create complex wave patterns to meet specific requirements. With our bench setup, this is not as readily accomplished.

We next modify the Timer 1 procedure—as shown in *Figure 4.31*—for operation in the single-shot mode. We want to keep the Timer 1 low-order counter input low in order to observe the response.

SHIFT REGISTER PROCEDURE FOR SHIFT OUT UNDER CONTROL OF CB1 AS A CLOCK

Note: Refer to the timing diagram of Figure 4.20(o).

Block the Port B Data Gate (Red LED ON)

1. Select the Auxiliary Control Register 11
   Mode 111 Shift Out under CB1 control
   RS3-RS0 = 1011 REG 11 SETUP ☐

OFF	
RSO	1
RS1	1
RS2	0
RS3	1
1 0	

7 6 5 4 3 2 1 0	OPERATION
0 0 0 1 1 1 0 0	Shift Out under CB1

   BUS DATA ENTRY 0001 1100 ☐
   SINGLE STEP CLOCK INPUT ☐

   REGISTER 11
   DIP SW SETTINGS

2. Select the Interrupt Enable Register 14
   RS3-RS0 = 1110 REG 14 SETUP ☐

OFF	
RSO	0
RS1	1
RS2	1
RS3	1
1 0	

7 6 5 4 3 2 1 0	OPERATION
1 0 0 0 0 1 0 0	Shift Register Interrupt

   BUS DATA ENTRY 1000 0100 ☐
   SINGLE STEP CLOCK INPUT ☐

   REGISTER 14
   DIP SW SETTINGS

3. Select the Shift Register to trigger the Shift Operation.
   RS3-RS0 = 1010 REG 10 SETUP ☐

OFF	
RSO	0
RS1	1
RS2	0
RS3	1
1 0	

7 6 5 4 3 2 1 0	OPERATION
Any data series	Write shift data in

   STEP ONE CLOCK CYCLE ONLY ☐

   REGISTER 10
   DIP SW SETTINGS

4. Select the Interrupt Flag Register 13.
   RS3-RS0 = 1101 REG 13 SETUP ☐

OFF	
RSO	1
RS1	0
RS2	1
RS3	1
1 0	

7 6 5 4 3 2 1 0	OPERATION
1 0 0 1 0 0 0 0	Complete 8 shifts

   BUS DATA DISPLAY 1001 0000 ☐
   Do NOT write to this register ☐

   REGISTER 13
   DIP SW SETTINGS

With the Clock Switch High, alternate CB1 DIP Switch position.
Toggle the Clock at least once. On first toggle IFR bits 4, 7
on while High, one cycle only.
On completion of the shift:
   IFR bits 4 and 7 ON ☐
   IRQ is Low ☐

CONTROL LINE DIP SWITCH SEQUENCE

STEP 1 ☐
OFF		
CB1		Open
CB2		O
CA1		1
CA2		1
1 0		

STEP 2 ☐
OFF		
CB1		Open
CB2		O
CA1		1
CA2		1
1 0		

STEP 3 ☐
OFF		
CB1		Open
CB2		O
CA1		1
CA2		1
1 0		

STEP 4 ☐
OFF		
CB1		Open
CB2		O
CA1		1
CA2		1

STEP 5 ☐
OFF		
CB1		Open
CB2		O
CA1		1
CA2		1
1 0		

STEP 6 ☐
OFF		
CB1		Open
CB2		O
CA1		1
CA2		1
1 0		

STEP 7 ☐
OFF		
CB1		Open
CB2		O
CA1		1
CA2		1
1 0		

STEP 8 ☐
OFF		
CB1		Open
CB2		O
CA1		1
CA2		1

Be sure that CB2 is disconnected from
its DIP switch
The 6522 CB1 pin 18 connection to
its DIP switch must be present.

O = 0 as shown.
Fill in the circle to indicate a one if
desired.
CA1, CA2 have no role to play.

5. Repeat steps 3 and 4 for ☐
   another cycle with new
   CB2 switch settings.

Figure 4.34. Procedure for setting up the shift register for shift-out under control of CB1 as the clock.

The seventh procedure, shown in *Figure 4.32*, is operation of Timer 2 in the one-shot mode. This procedure is very similar to that for Timer 1.

The remaining two procedures, shown in *Figures 4.33* and *4.34*, provide for data-shifting operations. Figure 4.33 sets up the shift register for shift-in operation under control of the Ø2 clock. Figure 4.34 provides for shift-out operation under control of CB1 as the clock.

The 6522 is very versatile, but also not the easiest device to operate on the bench. This is where persistence becomes a virtue. I have found the 6522 to be a challenging device to master. But I have also found satisfaction in the achievement. I hope you will also.

References

1. Synertek Systems Corporation, *VIM Reference Manual,* Appendix H: SY6522 Data Sheet, May 1978.

2. Synertek Systems Corporation, *SYM-1 Reference Manual,* 1980, Appendix J: SY6522 Data Sheet, April 1979.

3. Rockwell International, Electronics Division, *Data Catalog,* R6522 VIA, Document No. 2900 D47, Rev. 3, February 1981.

4. Marvin L. De Jong, *MICRO - The 6502 Journal,* "6522 Timing and Counting Techniques," October 1979.

5. Marvin L. De Jong, *MICRO - The 6502 Journal,* "Interfacing the 6522 Versatile Interface Adapter, January 1981.

6. The Western Design Center, W65C6522 Data Sheet, January 1998.

7. Reference for Figure 4.1: Synertek Systems Corporation, *SYM-1 Reference Manual,* 1980, Appendix J: SY6522 Data Sheet, April 1979, Figure 1.

8. Reference for Figure 4.2: Ibid. Tables, pp. 3, 4, Figures 3 and 4.

9. Reference for Figure 4.3: Ibid. Table, p. 5, Figures 5a-i, pp. 6-8.

10. Reference for Figure 4.4: Ibid. April 1979, Figure 9.

11. Reference for Figure 4.5: Ibid. Figure 10.

12. Reference for Figure 4.6: Ibid. Figure 11.

13. Reference for Figure 4.7: Ibid. Figures 7 and 8.

14. Reference for Figure 4.8: Ibid. Figure 14.

15. Reference for Figure 4.9: Ibid. Figure 13.

16. Reference for Figure 4.10: Ibid. Figure 15.

17. Reference for Figure 4.11: Ibid. Figure 15.

18. Reference for Figure 4.12: Ibid. Figure 16.

19. Reference for Figure 4.13: Ibid. Figure 16.

20. Reference for Figure 4.14: Ibid. Figures 18 and 19.

21. Reference for Figure 4.15: Ibid. Figure 20.

22. Reference for Figure 4.16: Ibid. Figure 20.

23. Reference for Figure 4.17: Ibid. Figure 21.

24. Reference for Figure 4.18: Ibid. Figure 22.

25. Reference for Figure 4.19: Ibid. Figure 22.

26. Reference for Figure 4.20: Ibid. Figures 23 and 24.

27. Reference for Figure 4.21: Ibid. Figure 24.

28. Reference for Figure 4.22: Ibid. Figure 25.

29. Reference for Figure 4.23: Ibid. Figure 26.

30. Reference for Table 4.2: Ibid. Figure 6.

Chapter 5
The R6520/ MC6820/
MC6821 PIA

Introduction

Chapter 4 provided details of the SY6522 versatile interface adapter, which is aptly named considering the many capabilities with which it has been endowed. The three devices that comprise the subject for this chapter are predecessors to the SY6522.

While they differ in some internal details, the three are functionally equivalent and interchangeable for many applications. Any of the three you happen to have on hand can be operated using the test layout and procedures for this chapter.

In many respects, we can view these three as a subset of the SY6522. The role of the PIA in the scheme of things is seen in the configuration drawing of *Figure 5.1*, where we see the manner in which it operates in a microcomputer system.

Figure 5.2 provides the package pinout and a functional block diagram. When we compare the pinout to that for the R6522, we see there are few changes made for these. The data bus and the two peripheral port connections are identical—as are the four controls, CA1, CA2, CB1, and CB2. Only two register selects—RS0 and RS1—are present. Enable (E) functions as the clock for the SY6522. There are two interrupt request pins: IRQA and IRQB.

Figure 5.3 defines the organization and functions of the control register. This figure also includes the register-addressing code structure. We will be returning to this drawing frequently as we progress with descriptions of the device features.

Figure 5.4 depicts the read and write timing diagrams. The value to us with these is the sequential ordering in device operations.

Operating times are not relevant to our slow-speed activities, but are available in the manufacturers literature. Device speeds of 1 MHz and 2 MHz are available. Response times are typically in the low nanosecond to microsecond range.

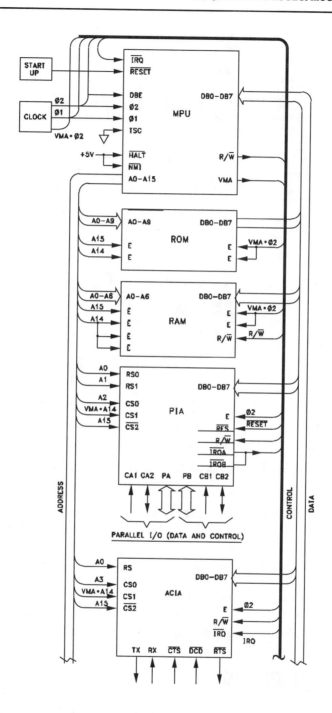

Figure 5.1. Microprocessor addressing example illustrating how the PIA relates to other members of the family.

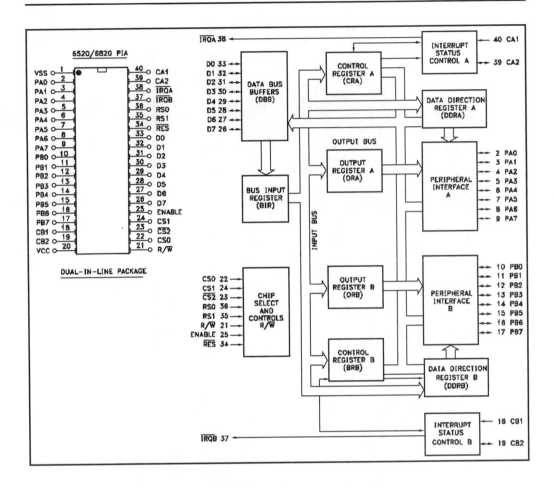

Figure 5.2. Expanded block diagram of the 6821 PIA. This diagram is also applicable to the 6520 and 6820 devices.

Figure 5.5 shows the nature of the port input and output circuitry in a simplified manner for the R6520 and the MC6820[1-4]. The MC6821 differs from these internally to some extent, yet there is no functional difference.

Theory of Operation

Frequent reference to Figures 5.1, 5.2, and 5.3 will be helpful while studying the signal functions that follow.

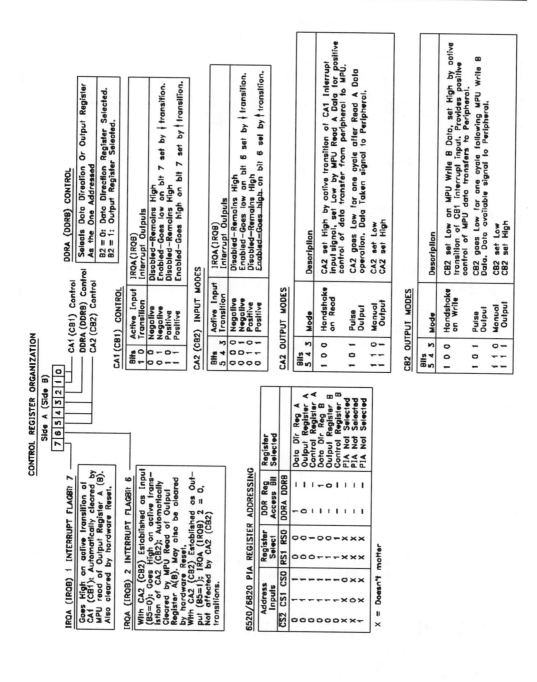

Figure 5.3.   Control register format of the 6520/6820/6821 PIA. The register-addressing table is included.

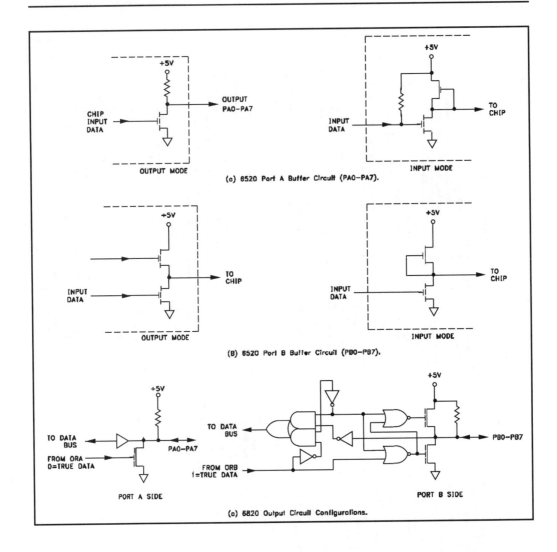

Figure 5.4A. The 6520 Port A buffer circuit (PAO-PA7). Figure 5.4B. The 6520 Port B buffer circuit (PBO-PB7). Figure 5.4C. The 6820 output-circuit configurations.

PIA to MPU Interface Signals

These consist of an eight-bit bidirectional data bus, three chip select lines, two register select lines, two interrupt request lines, a read/write line, an enable (clock) line, and a reset line.

READ TIMING CHARACTERISTICS (LOADING 130 pf AND ONE TTL LOAD)

Characteristics	Symbol	1 MHz Min	1 MHz Max	2 MHz Min	2 MHz Max	Unit
Delay Time, Address valid to Enable positive transition	tAEW	180	–	90	–	nS
Delay Time, Enable positive transition to Data valid on bus	tEDR	–	395	–	190	nS
Peripheral Data Setup Time	tPDSU	300	–	150	–	nS
Data Bus Hold Time	tHR	10	–	10	–	nS
Delay Time, Enable negative transition to CA2 negative transition	tCA2	–	1.0	–	0.5	uS
Delay Time, Enable negative transition to CA2 positive transition	tRS1	–	1.0	–	0.5	uS
Rise and Fall Time for CA1 and CA2 input signals	tr,tf	–	1.0	–	0.5	uS
Delay Time from CA1 active transition to to CA2 positive transition	tRS2	–	2.0	–	1.0	uS
Rise and Fall Time for Enable input	trE,tfE	–	25	–	25	nS

WRITE TIMING CHARACTERISTICS

Characteristics	Symbol	1 MHz Min	1 MHz Max	2 MHz Min	2 MHz Max	Unit
Enable Pulse Width	tE	0.470	25	0.235	25	uS
Delay Time, Address valid to Enable positive transition	tAEW	180	–	90	–	nS
Delay Time, Data valid to Enable negative transition	tDSU	300	–	150	–	nS
Delay Time, Read/Write negative transition to Enable positive transition	tWE	130	–	65	–	nS
Data Bus Hold Time	tHW	10	–	10	–	nS
Delay Time, Enable negative transition to Peripheral Data valid	tPDW	–	1.0	–	0.5	uS
Delay Time, Enable negative transition to Peripheral Data valid CMOS (Vcc–30%) PA0-PA7, CA2	tCMOS	–	2.0	–	1.0	uS
Delay Time, Enable positive transition to CB2 negative transition	tCB2	–	1.0	–	0.5	uS
Delay Time, Peripheral Data valid to CB2 negative transition	tDC	0	1.5	0	0.75	uS
Delay Time, Enable positive transition to CB2 positive transition	tRS1	–	1.0	–	0.5	uS
Rise and Fall Time for CB1 and CB2 input signals	tr,tf	–	1.0	–	0.5	uS
Delay Time, CB1 active transition to CB2 positive transition	tRS2	–	2.0	–	1.0	uS

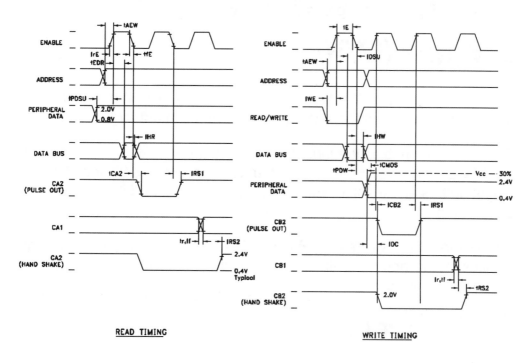

READ TIMING

WRITE TIMING

Figure 5.5.   Read and write timing characteristics and diagrams for the 6520 peripheral interface adapter.

The Bidirectional Data Bus (D0-D7)

In its system configuration, these lines provide for the transfer of data between the MPU and PIA. The bus drivers are three-state devices that remain in a high-impedance (off) state, except during a PIA read operation.

Enable (E)

The enable pulse is the only timing signal supplied to the PIA. Within a system, this is typically a clock pulse. Timing of all other signals is referenced to the leading and trailing edges of the E pulse. We can see this in the timing diagrams of *Figures 5.5, 5.6,* and *5.7.*

Read/Write (R//W)

This line controls the direction of flow on the data bus. When in the high state, the PIA mode is for reading the data bus content by the MPU.

The low state enables the input buffers for data transfer from the MPU to the PIA on the enable signal, if the device has been selected. The PIA output buffers are enabled when the proper address and the enable pulse are present.

Read/write timing is defined in Figure 5.5, which has relevance to all three devices. Read and write timing for the versions of MC6821 are further defined in Figure 5.6.

RESET (Active Low)

When low, this line resets all PIA registers to a logical zero (low). The PIA is reset at power on, or whenever this line is brought low.

BUS TIMING CHARACTERISTICS

Ident. No.	Characteristic	Symbol	MC6821		MC68A21		MC68B21		Unit
			Min	Max	Min	Max	Min	Max	
1	Cycle Time	tCYC	1.0	10	0.67	10	0.5	10	uS
2	Pulse Width, E Low	PWEL	430	–	280	–	210	–	nS
3	Pulse Width, E High	PWEH	450	–	280	–	220	–	nS
4	Clock Rise and Fall Time	tr,tf	–	25	–	25	–	20	nS
9	Address Hold Time	tAH	10	–	10	–	10	–	nS
13	Address Setup Time Before E	tAS	80	–	60	–	40	–	nS
14	Chip Select Setup Time Before E	tCS	80	–	60	–	40	–	nS
15	Chip Select Hold Time	tCH	10	–	10	–	10	–	nS
18	Read Data Hold Time	tDHR	20	50*	20	50*	20	50*	nS
21	Write Data Hold Time	tDHW	10	–	10	–	10	–	nS
30	Output Data Delay Time	tDDR	–	290	–	180	–	150	nS
31	Input Data Setup Time	tDSW	185	–	80	–	60	–	nS

Notes: 1. Voltage levels are VL≤ 0.4V, VH≥2.4V, unless otherwise specified.
2. Measurement points are 0.8V and 2.0V unless otherwise specified.

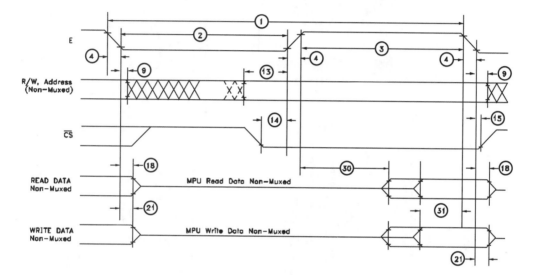

Figure 5.6. Bus timing characteristics and timing diagram for the MC6821/MC68A21/MC68B21 peripheral interface adapter.

Chip Selects (CS0, CS1, /CS2)

These three lines are used to select the current PIA in a system containing more than one. To be selected, CS0 and CS1 must be high and /CS2 must be low. The lines must be stable for the duration of the enable pulse.

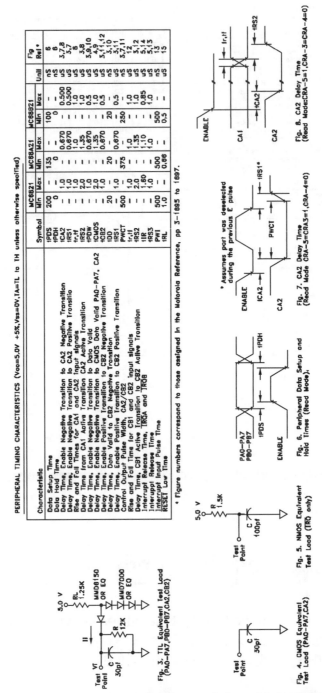

PERIPHERAL TIMING CHARACTERISTICS ($V_{DD}=5.0V \pm 5\%$, $V_{SS}=0V$, $T_A=T_L$ to T_H unless otherwise specified)

Figure 5.7. *(Continued on next page.)* **Peripheral timing characteristics and timing diagrams for the MC6821/MC68A21/MC68B21 PIA.**

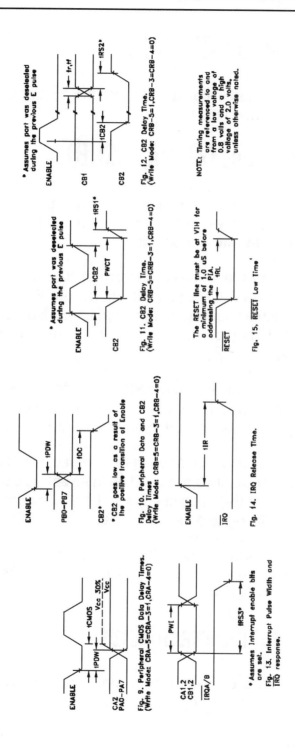

* Assumes port was deselected during the previous E pulse

Fig. 12. CB2 Delay Time.
(Write Mode: CRB-5=1,CRB-3=CRB-4=0)

NOTE: Timing measurements are referenced to and from a low voltage of 0.8 volts and a high voltage of 2.0 volts, unless otherwise noted.

* Assumes port was deselected during the previous E pulse

Fig. 11. CB2 Delay Time.
(Write Mode: CRB-5=CRB-3=1,CRB-4=0)

The RESET line must be at VIH for a minimum of 1.0 uS before addressing the PIA.

Fig. 15. RESET Low Time[1]

* CB2 goes low as a result of the positive transition of Enable

Fig. 10. Peripheral Data and CB2 Delay Times.
(Write Mode: CRB-5=CRB-3=1,CRB-4=0)

Fig. 14. IRQ Release Time.

Fig. 9. Peripheral CMOS Data Delay Times.
(Write Mode: CRA-5=CRA-3=1,CRA-4=0)

* Assumes interrupt enable bits are set.

Fig. 13. Interrupt Pulse Width and IRQ response.

Register Selects (RS0, RS1)

These two lines are used in conjunction with the control register for selection of the registers within the PIA.

Interrupt Request (/IRQA, /IRQB)

These lines, active when low, are used to interrupt the MPU either directly or through interrupt-priority circuitry. They are open drain, as we saw with the SY6522, to enable wire-or operation.

Each request line has two internal interrupt flag bits (refer to Figure 5.3) that can cause the IRQ request line to go low. Each flag bit is associated with a particular peripheral interrupt line. Also, four interrupt enable bits in the PIA may be used to inhibit a particular interrupt from a peripheral device.

The interrupt flags are cleared (zeroed) by a read of the peripheral data register. After being cleared, the interrupt flag bit cannot be enabled to be set until the PIA is deselected during an enable pulse.

The E pulse is also used to condition the interrupt control lines (CA1, CA2, CB1, CB2). When these lines are used as interrupt inputs, at least one E pulse must occur from the inactive edge to the active edge of the interrupt input signal to condition the edge-sense network.

If the interrupt flag has been enabled and the edge-sense circuit has been properly conditioned, the interrupt flag will be set on the next active transition of the interrupt input pin.

PIA Peripheral Interface Lines

The PIA provides two eight-bit bidirectional data buses and four interrupt/control lines for interfacing to peripheral devices.

Port A Peripheral Data (PA0-PA7)

Each data line can be individually programmed to act as an input or output. This is done by setting a "1" in the corresponding data-direction register bit for those lines designated as outputs. A "0" in a bit of this register designates the line as an input. Note this is consistent with the SY6522.

During an MPU read, Port A operation data on port lines programmed to act as inputs appears directly on the corresponding data bus lines. The data will be read properly if the voltage on these lines is greater than 2.0V for a logic "1" output and less than 0.8V for a logic "0" output.

Loading these lines such that the output does not reach full output levels results in error in the data read. The internal pull-up resistor on these represents a maximum of 1.5 standard TTL loads.

Port B Peripheral Data (PB0-PB7)

Each data line can be individually programmed to act as an input or output in a manner similar to Port A. These lines have three-state capability, such as to enter a high-impedance state when the port line is used as an input.

The data in Output Register B will appear on the data lines programmed to act as outputs. A logical "1" written into the register will cause a "high" on the corresponding data line; a "0" causes a "low." The error induced in an MPU data read by data-line loading doesn't occur with Port B, as it is the register content rather than the actual line level that is read. In addition to being compatible with TTL standards, the port lines can source up to 1 mA at 1.5V to drive a transistor switch.

Interrupt Input (CA1 and CB1)

These are input-only lines that set the interrupt flags of the control register. The active transition for these signals is also programmed by the two control registers.

Peripheral Control (CA2)

This line can be programmed to act as an interrupt input or as a peripheral control output. As an output, the line is compatible with standard TTL. As an input, the internal pull-up resistor represents 1.5 standard TTL loads. Its function is programmed by Control Register A. CA1 and CA2 timing in relation to the enable pulse is shown in Figures 5.5 and 5.7.

Peripheral Control (CB2)

This line can also be programmed to act as an interrupt input or as a peripheral control output. As an input, the line offers a high input impedance and is compatible with standard TTL.

As an output, the line is compatible with standard TTL and can also be used as a source of 1 mA at 1.5V to directly drive the base of a transistor switch. This line is programmed by Control Register B. CB1 and CB2 timing in relation to the enable pulse is shown in Figures 5.5 and 5.7.

Internal Controls

Initialization

The active low /RESET zeros all the PIA registers. PA0-PA7, BB0-PB7, CA2, and CB2 are all set as inputs, and the interrupts are disabled. Thus, following a reset, the device must be reconfigured.

There are six locations within the device accessible to the MPU data bus: the two peripheral registers, the two data-direction registers, and the two control registers. The selection of these is determined by inputs RS0 and RS1, as shown in the register-addressing table included with Figure 5.3.

Port A-B Hardware Characteristics

The two I/O ports differ from each other in several respects. The 6821 "A" side is designed to drive CMOS logic to normal 30% to 70% levels. An internal pull-up remains connected in both the input and output modes. Thus, the "A" side requires a greater drive current than the "B" side. The R6520/MC6820 "A" sides conform to TTL level specifications.

The MC6821 "B" is provided with three-state NMOS buffer circuitry, which cannot pull up to CMOS levels without external resistors. The R6520/MC6820 are TTL compatible. All three devices are, however, capable of driving a Darlington transistor base with 1.0 mA at 1.5V. When the MC6821 comes out of reset, the A port appears as inputs with internal pull-ups. The B port input will float high or low, depending on the circuitry to which it is connected.

For all devices, the read for the A port is of the actual pin voltage; the B port read is of the output latch ahead of the actual pin. This makes the A port output read sensitive to overloading conditions.

Control Registers CRA and CRB

The control registers permit MPU control of the four peripheral control lines CA1, CA2, CB1, and CB2. The registers also allow enabling of the interrupt lines and monitoring of the status of the interrupt flags. Bits 0 through 5 may be written to when the proper chip and register select inputs are applied. These are found in Figure 5.3. Bits 6 and 7 are read only, modified as indicated on the control lines CA1, CA2, CB1, and CB2. The timing diagrams of Figure 5.7 define the status of CRA and CRB for a variety of operating conditions. You may wish to refer to this figure as you continue your reading in the sections that follow.

Data Direction Access Control Bit, CRA-2 and CRB-2

Bit 2 in each control register determines the selection of either a peripheral output register or the corresponding data-direction enabling register as determined by RS0 and RS1. A "1" in Bit 2 enables access to the peripheral interface register. A "0" causes the data-direction register to be addressed.

Interrupt Flags, CRA-6, CRA-7, CRB-6, CRB-7

These four flag bits are set by active transitions of signals on the four interrupt and peripheral control lines when these are programmed as inputs. These bits cannot be set directly by the MPU data bus; they are reset indirectly by a read peripheral data operation on the appropriate section.

Control of CA2 and CB2 Peripheral Control Lines CRA-3, CRA-4, CRA-5, CRB-3, CRB-4, CRB-5

Control register Bits 3, 4, and 5 are used to control the CA2 and CB2 peripheral control lines. These bits determine if the control line will be an interrupt input or an output control signal. If register bit CRA-5 (CRB-5) is low, CA2 (CB2) is an interrupt line similar to CA1 (CB1). When this bit is high, CA2 (CB2) becomes an output signal that may be used to control peripheral data transfers. When in the output mode, the loading characteristics reflect those of their associated peripheral port.

Control of CA1 and CB1 Interrupt Lines CRA-0, CRB-0, CRA-1, CRB-1

The two lowest-order bits are used to control the interrupt input lines CA1 and CB1. The "0" bits enable the MPU interrupt signals /IRQA and /IRQB, respectively. The "1" bits determine the active transition of the interrupt input signals CA1 and CB1.

Bench-Top Operation of the 6520/6820/6821

Figures 5.8 and *5.9* illustrate my own layout for bench-top operation of the 6520/6820/6821 peripheral interface adapters. Figure 5.8 is a planning drawing provided as an aid in our wiring of the test configuration shown in

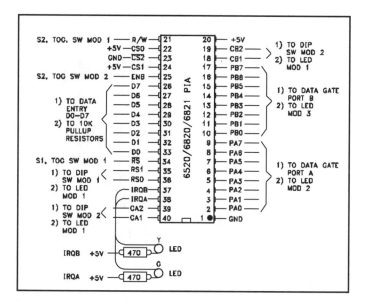

Figure 5.8. Wiring plan for the author's bench-top module layout and wiring configuration for operation of the 6520/6820/6821 PIA.

Figure 5.9. Here is shown the overall arrangement and the interconnecting wiring linking the modules and the PIA under test.

As noted previously, anyone of these three devices may be operated with the procedures that follow. Since your modules may appear differently from mine, the appearance of your setup may very well differ, but the functions must be provided.

I strongly recommend fastening your modules to a supporting base. I have found corrugated cardboard to be a cost-effective and a most satisfactory material for this. Just slice up a cardboard box having sides of an adequate size. Holes for attaching screws can be easily punched and the modules secured in place.

I use number 22 solid insulated wire of various colors and lengths (a practice of many years), so my stock is well organized by lengths. Do not be either too stingy or too generous with wire lengths. You want to be able to route the wiring around the switches and display LEDs without overdoing

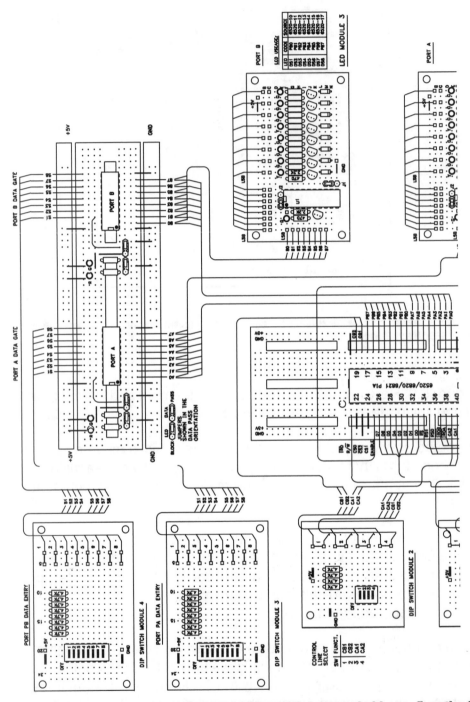

Figure 5.9. *(Continued on next page)* **The author's bench-top module layout and wiring configuration for operation of the 6520/6820/6821 PIA.**

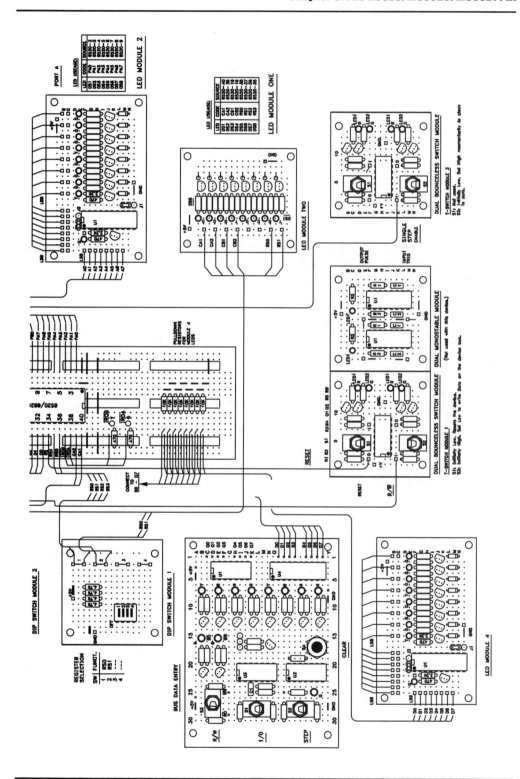

```
┌─────────────────────────────────────────────────────────────────┐
│  HOW TO USE THE DATA ENTRY AND 6520/6820/6821 R/W̄ TOGGLES        │
│  1. With the clock switch set LOW and both R/W̄ switches in the    │
│     Read position, set the data pattern on the Bus Data Entry Module. │
│     Observe that the pattern is correct on the Module output LEDs.│
│  2. Set the Register Selection DIP switches for the required register. │
│  3. To write to the Data Bus:                                     │
│     Set both R/W̄ switches to write.                               │
│     Note that the write pattern is present on LED Module 4.       │
│     Do a single clock step.                                       │
│  4. Return both R/W̄ switches to the Read position.                │
│  5. To verify that the write was successfully read do a single clock step. │
│     The data pattern written into the bus appears on LED Module 4 │
│     when the clock step is HIGH. The LEDs are blank with the clock LOW. │
└─────────────────────────────────────────────────────────────────┘
```

Table 5.1. Instructions on usage of the data-entry and R//W toggle switches found on the bus data-entry module and dual bounceless switch module.

it. As you can see, there is a substantial amount of wiring to be done. Take care with the connections; it is all too easy to plug into an adjacent point. Check yourself as you proceed. It's easier to catch mistakes now than later on when it will be very mysterious.

Review the switch usage procedures of *Table 5.1* carefully before beginning any operations. In particular, have an understanding of read and write as applied to these devices. If you have constructed the data-entry module as described in Chapter 1, you most likely have operated it sufficiently at this time to need no further instruction on its use. If not, now is a good time for a bit of practice.

If you are using a switching arrangement of your own devising, that is well and good—provided it meets the needs of entering data on the bus for write operations and isolating the bus when in the read mode. This requirement applies to the two peripheral ports as well. The port-gating circuitry for isolating the DIP switches is provided in Figure 1.11 of Chapter 1.

Figures 5.10 through *5.13* describe procedures for operating the device using the bench-top setup of Figure 5.9. These four procedures do not cover all the possible modes of the device operation. They will, however, provide insight into the register functions such that you can devise additional procedures of your own should these devices be of greater interest to you.

TO INPUT DATA ON PORT A:

CAUTION: Review Table 5.1 before beginning this procedure to ensure correct R/W switch and DIP switch useage.

1. Setup Control Register A for Port A as Input.
   Select DDRA as input (bit 2 = 0)

   RS1,RS0 = 01, CRA Setup ☐

   OFF
   RS0 [ ] 1
   RS1 [ ] 0
   -- [ ] --
   CS0 [ ] 1   ☐
       1 0
   REGISTER CRA
   DIP SW SETTINGS

7 6 5 4 3 2 1 0	OPERATION
0 0 0 0 0 0 0 0	DDRA set as Input

   BUS DATA ENTRY = 0000 0000 ☐
   SINGLE STEP CLOCK INPUT ☐

2. Select DDRA for Data Input
   RS1,RS0 = 00, DDRA Setup ☐

   OFF
   RS0 [ ] 0
   RS1 [ ] 0
   -- [ ] --
   CS0 [ ] 1   ☐
       1 0
   PORT A DIRECTION
   DIP SW SETTINGS

7 6 5 4 3 2 1 0	OPERATION
0 0 0 0 0 0 0 0	Port A set for Data Input

   BUS DATA ENTRY = 0000 0000 ☐
   SINGLE STEP CLOCK INPUT ☐

3. Set Control Register A for CA1, CA2 as shown.
   Select Port A for Data Input.
   RS1,RS0 = 01, CRA Setup ☐

   OFF
   RS0 [ ] 1
   RS1 [ ] 0
   -- [ ] --
   CS0 [ ] 1   ☐
       1 0
   REGISTER 3
   DIP SW SETTINGS

7 6 5 4 3 2 1 0	OPERATION
0 0 1 0 1 1 0 1	Port A set for Input CA1 Negative Edge CA2 Output Pulse

   BUS DATA ENTRY = 0010 1101 ☐
   SINGLE STEP CLOCK INPUT ☐
   CA2 LED, Module 1, OFF with Clock Low ☐

4. Select Port A for Data Entry. Read Port A.
   RS1,RS0 = 00, Read Port A ☐

   OFF
   RS0 [ ] 0
   RS1 [ ] 0
   -- [ ] --
   CS0 [ ] 1   ☐
       1 0
   READ PORT A
   DIP SW SETTINGS

7 6 5 4 3 2 1 0	OPERATION
	Input Data, port A

   BUS DATA ENTRY = (DIP SW) ☐
   CAUTION: Both R/W switches must be in R position.
   SINGLE STEP CLOCK INPUT ☐
   Note that data is displayed on LED Module 4 only while the clock is HIGH. DO NOT RESET THE CLOCK!

5. Select Control Register A to observe IRQA.
   RS1,RS0 = 01, CRA Setup ☐

   OFF
   RS0 [ ] 1
   RS1 [ ] 0
   -- [ ] --
   CS0 [ ] 1   ☐
       1 0
   REGISTER CRA
   DIP SW SETTINGS

7 6 5 4 3 2 1 0	OPERATION
1 0 1 0 1 1 0 1	Bit pattern with IRQA Flag

   OBSERVE FLAG OPERATION ☐
   SINGLE STEP CLOCK INPUT ☐
   Note: Bit 7 = 0 until Step 6 following.

Jumper the A Port Data Gate for source pass through. ☐

PORT A DIP SWITCH SETTING

Set CA1 = 1 ☐
Disconnect CA2 at DIP switch ☐

OFF
[ ] 1
[ ] 1
[ ] 0
[ ] 1
[ ] 1
[ ] 0
[ ] 1
[ ] 1
    1 0
PORT A DATA
ENTRY DIP SWITCH

OFF
CB1 [ ] 1
CB2 [ ] 1
CA1 [ ] 1
CA2 [ ] Open ☐
    1 0
CA, CB
SELECT
DIP SW

Note: It is instructive to change the data switch settings after each single step and observe the new data present on the Bus as seen on the LED Module 4 display.

NOTE:

The two R/W switches are to remain in the R position for steps 4 through 7. These steps may be repeated using new Port A DIP switch settings.

6. With Clock still High, CRA data displayed, set CA1 to 0. ☐
   Toggle DIP switch CS0 ☐
   IRQA goes Low ☐
7. Set RS0 = 0 ☐
   IRQA goes High ☐
8. Set CA1 = 1 ☐
   Toggle CS0 ☐
   CA2 LED, Module 1 ON with Clock Low ☐

Figure 5.10. Procedure for inputting data to peripheral Port A with CA1, CA2 participation in the data transfers.

TO INPUT DATA ON PORT B:

CAUTION: Review Table 5.1 before beginning this procedure to ensure correct R/W switch and DIP switch useage.

1. Setup Control Register B for Port B as Input
   Select DDRB as Input (bit 2 = 0)
   RS1,RS0 = 11, CRB Setup ☐

   OFF
   RSO ▢ 1
   RS1 ▢ 1
   -- ▢ -
   CSO ▢ 1 ☐
   1 0
   REGISTER CRB
   DIP SW SETTINGS

7 6 5 4 3 2 1 0	OPERATION
0 0 0 0 0 0 0 0	DDRB set as Input

   BUS DATA ENTRY = 0000 0000 ☐
   SINGLE STEP CLOCK INPUT ☐

2. Select DDRB for Data Input
   RS1,RS0 = 10, DDRB Setup ☐

   OFF
   RSO ▢ 0
   RS1 ▢ 1
   -- ▢ -
   CSO ▢ 1 ☐
   1 0
   PORT B DIRECTION
   DIP SW SETTINGS

7 6 5 4 3 2 1 0	OPERATION
0 0 0 0 0 0 0 0	Port B set for Data Input

   BUS DATA ENTRY = 0000 0000 ☐
   SINGLE STEP CLOCK INPUT ☐

3. Set Control Register B for CB1, CB2 as shown.
   Select Port B for Data Input.
   RS1,RS0 = 11, CRB Setup ☐

   OFF
   RSO ▢ 1
   RS1 ▢ 1
   -- ▢ -
   CSO ▢ 1 ☐
   1 0
   REGISTER CRB
   DIP SW SETTINGS

7 6 5 4 3 2 1 0	OPERATION
0 0 1 0 1 1 0 1	Port B set for Input CB1 Negative Edge CB2 Output Pulse

   BUS DATA ENTRY = 0010 1101 ☐
   SINGLE STEP CLOCK INPUT ☐
   CB2 LED, Module 1, ON with Clock Low ☐

4. Select Port B for Data Entry. Read Port B.
   RS1,RS0 = 10, Read Port B ☐

   OFF
   RSO ▢ 0
   RS1 ▢ 1
   -- ▢ -
   CSO ▢ 1 ☐
   1 0
   READ PORT B
   DIP SW SETTINGS

7 6 5 4 3 2 1 0	OPERATION
	Input Data, Port B

   BUS DATA ENTRY = (DIP SW) ☐
   CAUTION: Both R/W switches must be in R position.
   SINGLE STEP CLOCK INPUT ☐
   Note that data is displayed on LED Module 4 only while the clock is HIGH. DO NOT RESET THE CLOCK!

5. Select Control Register B to observe IRQB
   RS1,RS0 = 11, CRB Setup ☐

   OFF
   RSO ▢ 1
   RS1 ▢ 1
   -- ▢ -
   CSO ▢ 1 ☐
   1 0
   REGISTER CRB
   DIP SW SETTINGS

7 6 5 4 3 2 1 0	OPERATION
1 0 1 0 1 1 0 1	Bit pattern with IRQB Flag

   OBSERVE FLAG OPERATION ☐
   SINGLE STEP CLOCK INPUT ☐
   Note: Bit 7 = 0 until Step 6 following.

Jumper the B Port Data Gate for source pass through. ☐

PORT B DIP SWITCH SETTINGS
Set CB1 = 1 ☐
Disconnect CB2 of DIP switch ☐

OFF
▢ 0
▢ 1
▢ 1
▢ 0
▢ 1
▢ 0
▢ 1
▢ 1
1 0
PORT B DATA
ENTRY DIP SWITCH

OFF
CB1 ▢ 1
CB2 ▢ Open ☐
CA1 ▢ 1
CA2 ▢ 1
1 0
CA, CB
SELECT
DIP SW

Note: It is instructive to change the data switch settings after each single step and observe the new data present on the Bus as seen on the LED Module 4 display.

NOTE:

The two R/W switches are to remain in the R position for steps 4 through 7.
These steps may be repeated using new Port B DIP switch settings.

6. With Clock still High, CRB data displayed, set CB1 to 0. ☐
   Toggle DIP switch CSO ☐
   IRQB goes Low ☐
7. Set RS0 = 0 ☐
   IRQB goes High ☐
8. Set CB1 = 1 ☐
   Toggle CSO ☐
   CB2 LED, Module 1 ON with Clock Low ☐

Figure 5.11. Procedure for inputting data to peripheral Port B with CB1, CB2 participation in the data transfers.

TO OUTPUT DATA ON PORT A:

CAUTION: Review Table 4.5 befoer beginning this procedure to ensure correct R/W switch and DIP switch useage.

1. Setup Control Register A as Output.
   Ensure CA1 and CA2 both set to 1. ☐

   RS1,RSO = 01, CRA Setup ☐

   OFF
   RSO ☐ 1
   RS1 ☐ 0
   -- ☐ -
   CSO ☐ 1   ☐
     1 0
   REGISTER CRA
   DIP SW SETTINGS

7 6 5 4 3 2 1 0	OPERATION
0 0 0 0 0 0 0 0	Select DDRA

   BUS DATA ENTRY = 0000 0000 ☐
   SINGLE STEP CLOCK INPUT ☐

2. Set Port Register ORA for Output
   RS1,RSO = 00, ORA Setup

   OFF
   RSO ☐ 0
   RS1 ☐ 0
   -- ☐ -
   CSO ☐ 1   ☐
     1 0
   REGISTER ORA
   DIP SW SETTINGS

7 6 5 4 3 2 1 0	OPERATION
1 1 1 1 1 1 1 1	Set Port A for Data Output

   BUS DATA ENTRY = 1111 1111 ☐
   SINGLE STEP CLOCK INPUT ☐
   Module 2 LEDs Cleared on
   Clock return to Low ☐

3. Set Control Register A for CA1, CA2 as shown.
   RS1,RSO = 01, CRA Setup ☐

   OFF
   RSO ☐ 1
   RS1 ☐ 0
   -- ☐ -
   CSO ☐ 1   ☐
     1 0
   REGISTER CRA
   DIP SW SETTINGS

7 6 5 4 3 2 1 0	OPERATION
0 0 0 0 1 1 0 0	IRQA Interrupt Enable on CA2 active transition. (High-to-Low) Select Output Register A

   BUS DATA ENTRY = 0000 1100 ☐
   SINGLE STEP CLOCK INPUT ☐

4. Select Output Register ORA for Data Output. Write to Port A.
   RS1,RSO = 00, ORA Setup ☐

   OFF
   RSO ☐ 0
   RS1 ☐ 0
   -- ☐ -
   CSO ☐ 1   ☐
     1 0
   REGISTER ORA
   DIP SW SETTINGS

7 6 5 4 3 2 1 0	OPERATION
Any pattern	Output Data to Port A

   Place Data on Bus using Data Entry Module ☐
   SINGLE STEP CLOCK INPUT ☐
   Module 2 LEDs display data on clock's
   return to Low. ☐

5. Select Control Register A to observe IRQA.
   RS1,RSO = 01, CRA Setup ☐

   OFF
   RSO ☐ 1
   RS1 ☐ 0
   -- ☐ -
   CSO ☐ 1   ☐
     1 0
   REGISTER CRA
   DIP SW SETTINGS

7 6 5 4 3 2 1 0	OPERATION
0 1 0 0 1 1 0 0	Bit pattern with IRQA Low

   OBSERVE FLAG OPERATION ☐
   SINGLE STEP CLOCK INPUT ☐
   DO NOT LOWER THE CLOCK ☐
   R/W Switch remains in R
   Position.

Jumper the Port Data Gates
to block the DIP switch input. ☐

With both gates blocked:
Initially the Module 3 B LEDs OFF
         the Module 2 A LEDs ON

OFF
CB1 ☐ 1
CB2 ☐ 1
CA1 ☐ 1
CA2 ☐ 1   ☐
  1 0
CA, CB
SELECT
DIP SW

6. Toggle CSO DIP Switch ☐
   Set CA2 DIP Switch to 0 ☐
   IRQA goes Low(IRQA LED ON) ☐
7. Set CA2 = 1 ☐
   Set RSO = 0 ☐
   IRQA goes High (IRQA LED OFF) ☐
8. Set RSO = 1 ☐
   Flag Bit 6 LED is OFF ☐

Figure 5.12. Procedure for outputting data from peripheral Port A with CA1, CA2 participation in the data transfers.

TO OUTPUT DATA ON PORT B:

CAUTION: Review Table 4.5 befoer beginning this procedure to ensure correct R/W switch and DIP switch useage.

1. Setup Control Register B as Output.

Ensure CB1 and CB2 both set to 1. ☐

RS1,RS0 = 11, CRB Setup ☐

OFF			7 6 5 4 3 2 1 0	OPERATION
RSO		1	0 0 0 0 0 0 0 0	Select DDRB
RS1		1		
--		-		
CSO		1		

1 0
REGISTER CRB
DIP SW SETTINGS

BUS DATA ENTRY = 0000 0000 ☐
SINGLE STEP CLOCK INPUT ☐

2. Set Port Register ORB forOutput.

RS1,RS0 = 10, ORB Setup

OFF			7 6 5 4 3 2 1 0	OPERATION
RSO		0	1 1 1 1 1 1 1 1	Set Port B for Data Output
RS1		1		
--		-		
CSO		1		

1 0
REGISTER ORB
DIP SW SETTINGS

BUS DATA ENTRY = 1111 1111 ☐
SINGLE STEP CLOCK INPUT ☐

3. Set Control Register B for CB1, CB2 as shown

RS1,RS0 = CRB Setup ☐

OFF			7 6 5 4 3 2 1 0	OPERATION
RSO		1	0 0 0 0 1 1 0 0	IRQB Interrupt Enable on CB2 active transition (High-to-Low) Select Output Register B
RS1		1		
--		-		
CSO		1		

1 0
REGISTER CRB
DIP SW SETTINGS

BUS DATA ENTRY = 0000 1100 ☐
SINGLE STEP CLOCK INPUT ☐

4. Seleat Output Register ORB for Data Output. Write to Port B.

RS1,RS0 = 10, ORB Setup ☐

OFF			7 6 5 4 3 2 1 0	OPERATION
RSO		0	Any pattern	Output Data to Port B
RS1		1		
--		-		
CSO		1		

1 0
REGISTER ORB
DIP SW SETTINGS

Place Data on Bus using Data Entry Module ☐
SINGLE STEP CLOCK INPUT ☐
Module 2 LEDs display data on clock's return to Low. ☐

5. Select Control Register B to observe IRQB.

RS1,RS0 = 11, CRB Setup ☐

OFF			7 6 5 4 3 2 1 0	OPERATION
RSO		1	0 1 0 0 1 1 0 0	Bit pattern with IRQB Low
RS1		1		
--		-		
CSO		1		

1 0
REGISTER CRB
DIP SW SETTINGS

OBSERVE FLAG OPERATION ☐
SINGLE STEP CLOCK INPUT ☐
DO NOT LOWER THE CLOCK ☐
R/W Switch remains in R Position. ☐

Jumper the Port Data Gates to block the DIP switch input. ☐

With both gates blocked:
Initially the Module 3 B LEDs OFF
the Module 2 A LEDs ON

OFF		
CB1		1
CB2		1
CA1		1
CA2		1	☐

1 0
CA, CB
SELECT
DIP SW

6. Toggle CSO DIP Switch ☐
Set CB2 DIP Switch to 0 ☐
IRQB goes Low (IRQB LED ON) ☐

7. Set CB2 = 1 ☐
Set RSO = 0 ☐
IRQB goes High (IRQB LED OFF) ☐

8. Set RSO = 1 ☐
Flag Bit 6 LED is OFF ☐

Figure 5.13. Procedure for outputting data from peripheral Port B with CB1, CB2 participation in the data transfers.

The first procedure, shown in Figure 5.10, provides for inputting data to peripheral port A with CA1 and CA2 participating in the transfers. Each procedure is designed to be self-explanatory. Even so, it will be helpful to review the text on related features and the procedures with care before and/or during your operation of the device.

The next procedure, shown in *Figure 5.11*, is for inputting data to peripheral port B utilizing CB1 and CB2. You will see that it is very similar to the preceding procedure. *Figure 5.12* is a procedure for outputting data from peripheral port A with the participation of CA1 and CA2 in the transfer. *Figure 5.13* is a similar procedure for outputting data from peripheral port B with CB1 and CB2 participating in the transfer. These devices are less versatile than the 6522, thus you may find these to be less challenging devices to master than the 6522. But I have found there is still satisfaction to be had in the achievement. I hope you will find this true for yourself also.

References

1. Synertek Systems Corporation, *SY6520/SY6520A, SY6820/SY6820A Peripheral Interface Adapter (PIA)* Data Sheet

2. Rockwell International, Electronics Devices Division, *Data Catalog,* R6520 Data Sheet, 1978

3. Motorola, Inc., *Microprocessor, Microcontroller and Peripheral Data, Volume II,* MC6821 Peripheral Interface Adapter (PIA), 1988

4. Motorola Semiconductor Products, Inc., *Microprocessor Applications Manual,* Section 3.4 "Program Controlled Data Transfers," Section 3.4.1 "MC6820 Peripheral Interface Adapter, 1975

5. Reference for Figure 5.1: Motorola Semiconductor Products, Inc., *Microprocessor Applications Manual,* 1975, Section 3.4 "Program Controlled Data Transfers," Figure 3-4.1.3-2.

6. Reference for Figure 5.2: Motorola Semiconductor Products, Inc., *Microprocessor Applications Manual,* 1975, Section 3.4 "Program Controlled Data Transfers," Figure 3-4.1.2-1.

7. References for Figure 5.3:
   1. Motorola Semiconductor Products, Inc., *Microprocessor Applications Manual,* 1975, Section 3.4 "Program Controlled Data Transfers," Figure 3-4.1.2-3.
   2. Rockwell International, Electronics Devices Division, *Data Catalog,* R6520 Data Sheet, 1978.

8. References for Figure 5.4:

   1. Synertek Systems Corporation, *SY6520/SY6520A, SY6820/ SY6820A Peripheral Interface Adapter (PIA)* Data Sheet. (Figure 5.1a,b)

   2. Motorola Semiconductor Products, Inc., *Microprocessor Applications Manual,* 1975, Section 3.4 "Program Controlled Data Transfers," Figure 3-4.1.2-2. (Figure 5.1c)

9. Reference for Figure 5.5: Rockwell International, Electronics Devices Division, *Data Catalog,* R6520 Data Sheet, 1978.

10. Reference for Figure 5.6: Motorola, Inc., *Microprocessor, Microcontroller and Peripheral Data, Volume II,* MC6821 Peripheral Interface Adapter (PIA), 1988. Bus Timing Characteristics and Figure 1, Bus Timing, p 3-1694.

11. Reference for Figure 5.7: Motorola, Inc., *Microprocessor, Microcontroller and Peripheral Data, Volume II,* MC6821 Peripheral Interface Adapter (PIA), 1988. Peripheral Timing Characteristics and Figures 3-15, Peripheral Timing, pp 3-1695 to 3-1697.

Chapter 6
The INS8250, INS8250-B, NS16450, INS8250A, NS16C450, and INS82C50A UARTs

Introduction

Chapter 3 described the TR1602/AY-5-1013, a UART meeting the basic requirements for serial data transmission. The devices we will learn about in this chapter provide this same function but with additional capabilities, including a programmable baud-rate generator and complete control of modem operations.

Earlier versions of the 8250-B were described as an asynchronous communication element (ACE). This device differs in certain respects from the A version and the 16450 series. The differences are described later in the chapter.

References in this chapter are simply to the *generic* 8250[1], unless a distinction between the INS8250A[2]/ NS16450 or the 8250-B requires clarification. The NS16450 is a version of the 8250A with improved performance specifications. The NS16C450 and INS82C50 are functionally equivalent to their XMOS TTL-level counterparts, except that they are CMOS parts.

Figure 6.1 illustrates the basic MICROBUS[3] configuration for the series.

Two INS8250A/NS16450 packages are available: the 40-pin DIP and a 44-pin PCC. The DIP package of *Figure 6.2* and the block diagram show the 8250A/NS16450 pin designations with the alternative, early 8250-B defined for reference.

The current INS8250-B designations provided in the reference are the same as for the INS8250A/NS16450 and are used throughout the chapter. The function diagrams apply equally well to all members of this device family.

Figures 6.3 and *6.4* illustrate applications for the two versions. From these, we can see the similarity in their usage. These diagrams will be helpful to our understanding as we perform the bench-top operations later.

The basic functions of the device are: (1) to accept parallel input from the data bus, perform a serial-to-parallel conversion and output it in serial format, and (2) to receive serial input and to present it to the CPU bus in a

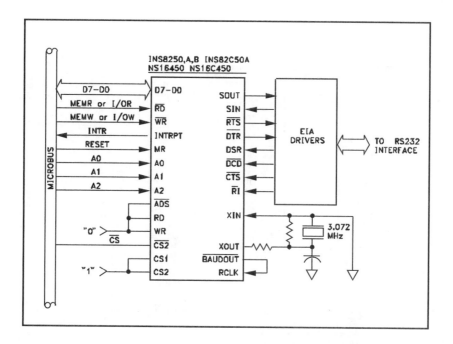

Figure 6.1. The MICROBUS connection diagram. Though some designations shown apply to the 8250-B, the diagram is applicable to all devices in this chapter.

parallel format. A wide range of baud rates is available internally, as are the communication protocols required for modem operation.

As we see in the application drawings, the clock source may be linked to that for the system. There is also an alternative crystal source option seen in *Figure 6.5*. The baud rates obtainable using one of two crystal frequencies are described in *Tables 6.3a* and *6.3b*. Related BAUDOUT waveforms are illustrated in *Figure 6.6*.

Timing diagrams for read and write cycles are shown in *Figure 6.7*. This figure includes the AC characteristics for the 8250-B. This device is somewhat slower than the 8250A series. The data is shown for reference only as speed with our bench-top operation is irrelevant.

The diagrams themselves are helpful, however, in observing the operational sequences required. If timing data is important to your needs, the manufacturer's data should be consulted.

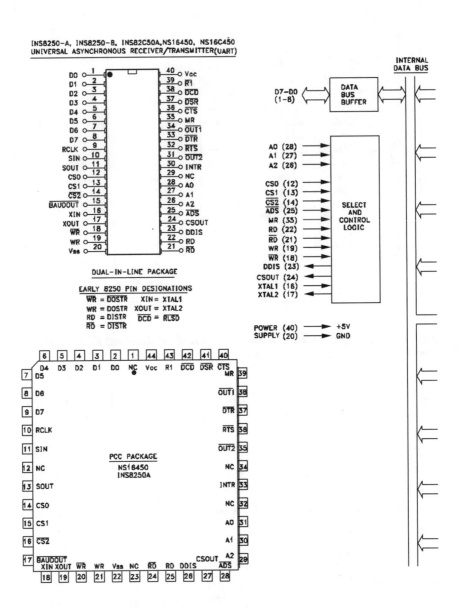

Figure 6.2. *(Continued on next page.)* **Package pinouts applicable as shown and the block diagram applicable to all devices in this chapter. Early INS8250-B data sheets show the alternative pin designations.**

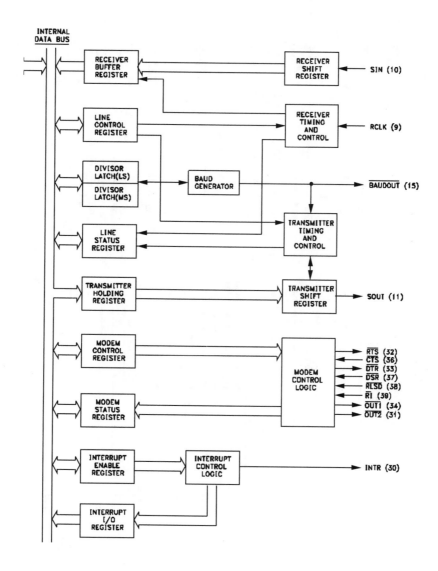

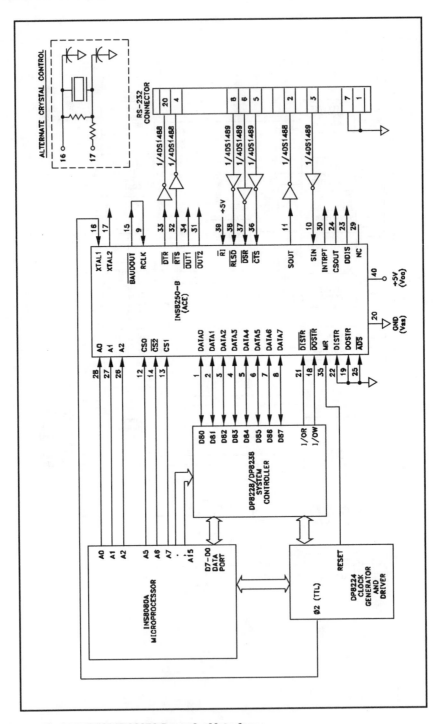

Figure 6.3. Typical INS8080A/INS8250-B terminal interface.

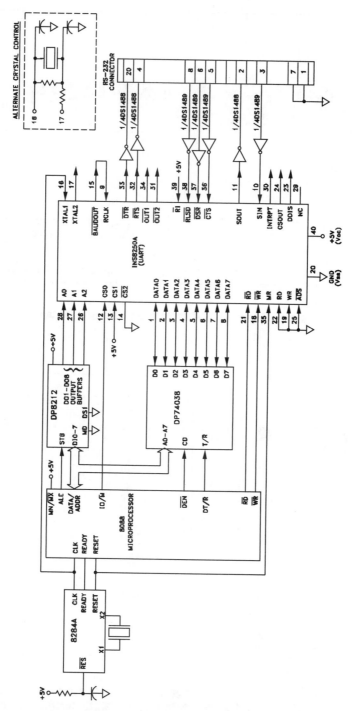

Figure 6.4. Basic connections of an INS8250A to an 8088 CPU.

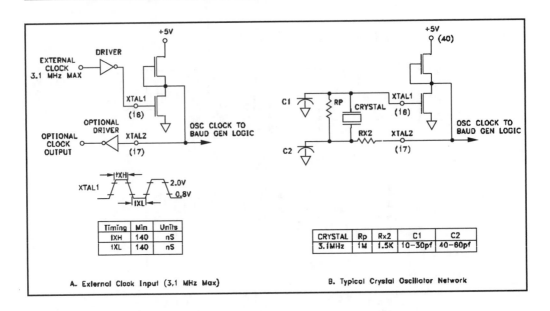

A. External Clock Input (3.1 MHz Max)

B. Typical Crystal Oscillator Network

Timing	Min	Units
tXH	140	nS
tXL	140	nS

CRYSTAL	Rp	Rx2	C1	C2
3.1MHz	1M	1.5K	10–30pf	40–60pf

Figure 6.5. Clock applications: (A) connection diagram for an external clock source; (B) a typical crystal oscillator network.

BAUD RATES USING 1.8432 MHz CRYSTAL

Desired Baud Rate	Divisor Used to Generate 16x Clock	Percent Error Difference Between Desired & Actual
50	2304	—
75	1536	—
110	1047	0.026
134.5	857	0.058
150	768	—
300	384	—
600	192	—
1200	96	—
1800	64	—
2000	58	0.69
2400	48	—
3600	32	—
4800	24	—
7200	16	—
9600	12	—
19200	6	—
38400	3	—
56000	2	2.86

NOTE: 1.8432 MHz is the standard 8080 frequency divided by 10.

BAUD RATES USING 3.072 MHz CRYSTAL

Desired Baud Rate	Divisor Used to Generate 16x Clock	Percent Error Difference Between Desired & Actual
50	3840	—
75	2560	—
110	1745	0.026
134.5	1428	0.034
150	1280	—
300	640	—
600	320	—
1200	160	—
1800	107	0.312
2000	96	—
2400	80	—
3600	53	0.628
4800	40	—
7200	27	1.23
9600	20	—
19200	10	—
38400	5	—

Table 6.3. Baud rates obtained by divisors used to generate 16X clocks using a crystal frequency of 1.8432 MHz, and a crystal frequency of 3.072 MHz. Coordinate this table with the text provided under the heading "Programmable Baud Generator."

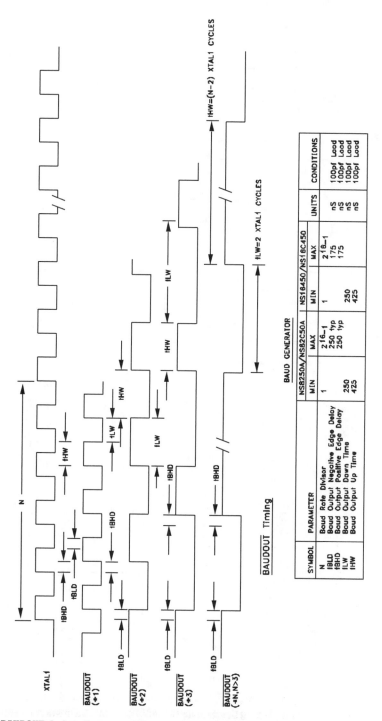

Figure 6.6. BAUDOUT timing diagram with generator representative times.

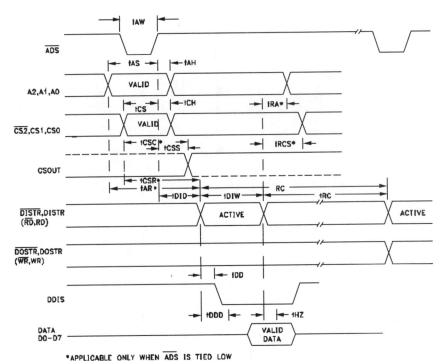

READ CYCLE

INS8250-B AC

SYMBOL	PARAMETER	MIN	MAX	UNITS	CONDITIONS
tAW	Address Strobe Width	120	–	nS	
tAS	Address Setup Time	110	–	nS	
tAH	Address Hold Time	60	–	nS	
tCS	Chip Select Setup Time	110	–	nS	
tCH	Chip Select Hold Time	60	–	nS	
tCSS	Chip Select Output Delay from Strobe	0	100	nS	100 pf Loading
tDID	DISTR/DISTR Strobe Delay	0	–	nS	
tDIW	DISTR/DISTR Strobe Width	350	–	nS	
tRC	Read Cycle Delay	1735	–	nS	
RC	Read Cycle = tAW+tDID+tDIW+tRC	2205	–	nS	
tDD	DISTR/DISTR to Driver Disable Delay	–	250	nS	100 pf Loading
tDDD	Delay from DISTR/DISTR to Data	–	300	nS	100 pf Loading
tHZ	DISTR/DISTR to Floating Data Delay	100	–	nS	100 pf Loading
tDOD	DOSTR/DOSTR Strobe Delay	50	–	nS	
tDOW	DOSTR/DOSTR Strobe Width	350	–	nS	

NS16450/NS16C450

SYMBOL	PARAMETER	MIN	MAX	UNITS	CONDITIONS
tADS	Address Strobe Width	60		nS	
tAH	Address Hold Time	0		nS	
tAR	RD,RD Delay from Address	60		nS	Note 1
tAS	Address Setup Time	60		nS	
tAW	WR,WR Delay from Address	60		nS	Note 1
tCH	Chip Select Hold Time	0		nS	
tCS	Chips Select Setup Time	60		nS	
tCSC	Chip Select Output Delay from Select		100	nS	@100pf Ldg, N1
tCSR	RD,RD Delay from Chip Select	50		nS	Note 1
tCSW	WR,WR Delay from Select	50		nS	Note 1
tDH	Data Hold Time	40		nS	
tDS	Data Setup Time	40		nS	
tHZ	RD,RD to Floating Data Delay	0	100	nS	@100pf Ldg, N3
tMR	Master Reset Pulse Width	5		nS	
tRA	Address Hold Time from RD,RD	20		nS	Note 1

Figure 6.7. *(Continued on next page.)* **Timing diagram for the INS8250-B and NS16450 read and write cycles with AC characteristics given.**

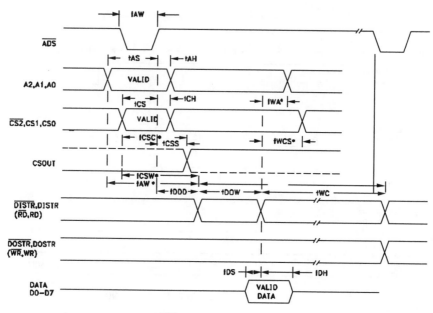

*APPLICABLE ONLY WHEN ADS IS TIED LOW

CHARACTERISTICS

WRITE CYCLE

SYMBOL	PARAMETER	MIN	MAX	UNITS	CONDITIONS
tWC	Write Cycle Delay	1785	–	nS	
WC	Write Cycle = tAW+tDOD+tDOW+tWC	2305		nS	
tDS	Data Setup Time	350	–	nS	
tDH	Data Hold Time	100	–	nS	
tCSC*	Chip Select Output Delay from Select	–	200	nS	100 pf Loading
tRA*	Address Hold Time from DISTR/DISTR	50	–	nS	
tRCS*	Chip Select Hold Time from DISTR/DISTR	50	–	nS	
tAR*	DISTR/DISTR Delay from Address	110	–	nS	
tCSR*	DISTR/DISTR Delay from Chip Select	110	–	nS	
tWA*	Address Hold Time from DOSTR/DOSTR	50	–	nS	
tWCS*	Chip Select Hold Time from DOSTR/DOSTR	50	–	nS	
tAW*	DOSTR/DOSTR Delay from Address	160	–	nS	
tCSW*	DOSTR/DOSTR Delay from Select	160	–	nS	
tMRW	Master Reset Pulse Width	25	–	uS	

AC CHARACTERISTICS

SYMBOL	PARAMETER	MIN	MAX	UNITS	CONDITIONS
tRC	Read Cycle Delay	175		nS	
tRCS	Chip Select Hold Time from RD,RD	20		nS	Note 1
tRD	RD,RD Strobe Width	125		nS	
tRDD	RD,RD to Driver Disable Delay		60	nS	@100pf Ldg, N3
tRVD	Delay from RD,RD to Data		125	nS	@100pf Ldg
tWA	Address Hold Time from WR,WR	20		nS	Note 1
tWC	Write Cycle Delay	200		nS	
tWCS	Chip Select Hold Time from WR,WR	20		nS	Note 1
tWR	WR,WR Strobe Width	100		nS	
tXH	Duration of Clock High Pulse	140		nS	
tXL	Duration of Clock Low Pulse	140		nS	
RC	Read Cycle=tAR+tRD+tRC	360		nS	Ext. Clk, 3.1MHz
WC	Write Cycle=tAW+tWR+TWC	360		nS	Ext. Clk, 3.1MHz

NOTES: 1. Applicable only when ADS is tied low.
2. RCLK is equal to tXH and tXL.
3. Charge and discharge time is determined by VOL, VOH and the external loading.

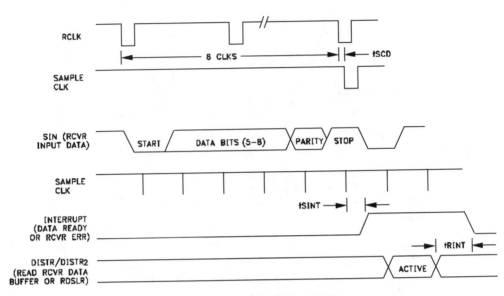

RECEIVER TIMING

INS8250-B AC ELECTRICAL CHARACTERISTICS

SYMBOL	PARAMETER	MIN	MAX	UNITS	CONDITIONS
RECEIVER					
tSCD	Delay from RCLK to Sample Time	–	2 typ	uS	100 pf Load
tSINT	Delay from Stop to Set Interrupt	–	2 typ	uS	100 pf Load
tRINT	Delay from DOSTR/DOSTR (RD RDR/RDLSR) to Reset Interrupt	–	1 typ	uS	
TRANSMITTER					
tHR	Delay from DOSTR/DOSTR (WR THR) to Reset Interrupt	–	1 typ	uS	100 pf Load
tIRS	Delay from Initial INTR Reset to Transmit Start	–	16 typ	BAUDOUT Cycles	
tSI	Delay from Initial Write to Interrupt	–	24 typ	BAUDOUT Cycles	
tSS	Delay from Stop to next Start	–		uS	
tSTI	Delay from Stop to Interrupt (THRE)	–	8 typ	BAUDOUT Cycles	
tIR	Delay from DISTR/DISTR (RD IIR) to Reset Interrupt	–	1 typ	uS	100 pf Load

Figure 6.8. *(Continued on next page.)* **Timing diagram for the INS8250-B and NS16450 receiver and transmitter timing with AC characteristics given.**

Timing diagrams for receiver and transmitter functions are shown in *Figure 6.8*. Timing diagrams for modem controls are shown in *Figure 6.9*. Figures 6.8 and 6.9 also include the AC characteristics for the 8250-B. The remaining figures relate to the bench-top operation and are described in that section of the chapter.

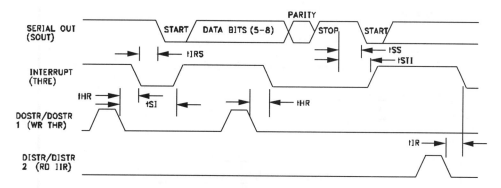

TRANSMITTER TIMING

NS16450/NS16C450 AC CHARACTERISTICS

SYMBOL	PARAMETER	MIN	MAX	UNITS	CONDITIONS
RECEIVER					
tSCD	Delay from RCLK to Sample Time		2	uS	
tSINT	Delay from Stop to Set Interrupt		1	RCLK Cycles (Note 2)	
tRINT	Delay from RD,RD (RD RBR or RD LSR) to Reset Interrupt		1	uS	100 pf Load
TRANSMITTER					
tHR	Delay from WR,WR (WR THR) to Reset Interrupt		175	nS	100 pf Load
tIR	Delay from RD,RD (RD IIR) to Reset Interrupt (THRE)		250	nS	100 pf Load
tIRS	Delay from initial INTR Reset to Transmit Start	24	40	BAUDOUT Cycles	
tSI	Delay from Initial Write to Interrupt	16	24	BAUDOUT Cycles	Note 1
tSTI	Delay from Stop to Interrupt (THRE)	8	8	BAUDOUT Cycles	

Notes: 1. For NS16C450 and INS82C50A, tSI is a minimum of 16 and a maximum of 48 BAUDOUT cycles.
2. RCLK is equal to tXH and tXL.

Table 6.1 relates the address inputs to the register they select. We find 10 registers residing within the block diagram of Figure 6.2. Table 6.1 defines the addressing for these and the divisor latch access bit (DLAB).

Table 6.2 describes the master reset functions, about which we will learn more later.

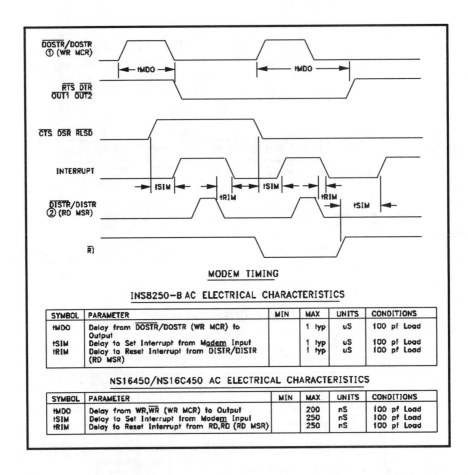

MODEM TIMING

INS8250-B AC ELECTRICAL CHARACTERISTICS

SYMBOL	PARAMETER	MIN	MAX	UNITS	CONDITIONS
tMDO	Delay from DOSTR/DOSTR (WR MCR) to Output		1 typ	uS	100 pf Load
tSIM	Delay to Set Interrupt from Modem Input		1 typ	uS	100 pf Load
tRIM	Delay to Reset Interrupt from DISTR/DISTR (RD MSR)		1 typ	uS	100 pf Load

NS16450/NS16C450 AC ELECTRICAL CHARACTERISTICS

SYMBOL	PARAMETER	MIN	MAX	UNITS	CONDITIONS
tMDO	Delay from WR,WR (WR MCR) to Output		200	nS	100 pf Load
tSIM	Delay to Set Interrupt from Modem Input		250	nS	100 pf Load
tRIM	Delay to Reset Interrupt from RD,RD (RD MSR)		250	nS	100 pf Load

Figure 6.9. Diagram for the 8250-B and NS16450 modem timing with AC characteristics given.

Tables 6.3a and 6.3b define available baud rates with a 1.8432 MHz and 3.072 MHz crystal, respectively.

Table 6.4 provides a summary of the register addressing and their functions. These registers control all of the device operations, including the transmission and reception of data. The registers are accessible from the CPU for their operation, which, in this book, is *us*!

Table 6.5 shows the four-bit code selection for the word length range of five to eight bits. *Table 6.6* provides the addressing and function descriptions for the interrupt control functions.

REGISTER SELECT TABLE

DLAB	A2	A1	A0	REGISTER
0	0	0	0	Receiver Buffer (Read)
				Transmitter Buffer (Write)
0	0	0	1	Interrupt Enable
X	0	1	0	Interrupt ID (Read Only)
X	0	1	1	Line Control
X	1	0	0	MODEM Control
X	1	0	1	Line Status
X	1	1	0	MODEM Status
X	1	1	1	None
1	0	0	0	Divisor Latch (LSB)
1	0	0	1	Divisor Latch (MSB)

Table 6.1. *(At left)* **The DLAB register select addressing. Note that all of these registers are contained within the divisor latch access and communicate with the CPU via the eight-bit data bus. The DLAB is described in Table 6.4.**

REGISTER/SIGNAL	RESET CONTROL	RESET STATE
Interrupt Enable Register	Master Reset	All Bits Low
		(0-3 Forced, 4-7 Permanent)
Interrupt Identification Register	Master Reset	Bit 0 is High, Bits 1 and 2 Low
		Bits 3-7 are Permanently Low
Line Control Register	Master Reset	All Bits Low
MODEM Control Register	Master Reset	All Bits Low
Line Status Register	Master Reset	All Bits Low
		Except Bits 5,6 are High
MODEM Status Register	Master Reset	Bits 0-3 Low
		Bits 4-7 — Input Signal
SOUT	Master Reset	High
INTRPT (RCVR Errs)	Read LSR/MR	Low
INTRPT (RCVR Data Ready)	Read RBR/MR	Low
INTRPT (THRE)	Read IIR/Write THR/MR	Low
INTRPT (Modem Status Changes)	Read MSR/MR	Low
OUT2	Master Reset	High
RTS	Master Reset	High
DTR	Master Reset	High
OUT1	Master Reset	High

Table 6.2. *(Above)* **UART reset functions. Observe that each DLAB register is unique with respect to the master reset. Coordinate this table with the master reset definitions provided under the heading "UART/ ACE Pin Description".**

Table 6.5. *(At right)* **This table defines the DLAB bit coding for the data word length as coded for serial transmission. Note that there is a relationship between word length and the number of stop bits.**

LINE CONTROL REGISTER

BIT 1	BIT 0	WORD LENGTH
0	0	5 Bits
0	1	6 Bits
1	0	7 Bits
1	1	8 Bits

The remaining tables relate to the bench-top operation and are described in that section of the chapter.

SUMMARY OF REGISTERS

REGISTER ADDRESS

Bit No.	0 DLAB = 0 Receiver Buffer Register (Read Only) RBR	0 DLAB = 0 Transmitter Holding Register (Write Only) THR	1 DLAB = 0 Interrupt Enable Register IER	2 Interrupt Identification Register (Read Only) IIR	3 Line Control Register LCR	4 MODEM Control Register MCR	5 Line Status Register LSR	6 MODEM Status Register MSR	0 DLAB = 1 Divisor Latch (LS) DLL	1 DLAB = 1 Divisor Latch (MS) DLM
0	Data Bit 0*	Data Bit 0	Enable Received Data Available Interrupt (ERBFI)	"0" if Interrupt Pending	Word Length Select Bit 0 (WLS0)	Data Terminal Ready (DTR)	Data Ready (DR)	Delta Clear to Send (DCTS)	Bit 0	Bit 8
1	Data Bit 1	Data Bit 1	Enable Transmitter Holding Register Empty Interrupt (ETBEI)	Interrupt ID Bit (0)	Word Length Select Bit 1 (WLS1)	Request to Send (RTS)	Overrun Error (OR)	Delta Data Set Ready (DDSR)	Bit 1	Bit 9
2	Data Bit 2	Data Bit 2	Enable Receiver Line Status Interrupt (ELSI)	Interrupt ID Bit (1)	Number of Stop Bits (STB)	OUT 1	Parity Error (PE)	Trailing Edge Ring Indicator (TERI)	Bit 2	Bit 10
3	Data Bit 3	Data Bit 3	Enable MODEM Status Interrupt (EDSSI)	0	Parity Enable (PEN)	OUT 2	Framing Error (FE)	Delta Receiver Line Signal Detect	Bit 3	Bit 11
4	Data Bit 4	Data Bit 4	0	0	Even Parity Select (EPS)	Loop	Break Interrupt (BI)	Clear To Send (CTS)	Bit 4	Bit 12
5	Data Bit 5	Data Bit 5	0	0	Stick Parity	0	Transmitter Holding Register Empty (THRE)	Data Set Ready (DSR)	Bit 5	Bit 13
6	Data Bit 6	Data Bit 6	0	0	Set Break	0	Transmitter Shift Register Empty (TSRE)	Ring Indicator (RI)	Bit 6	Bit 14
7	Data Bit 7	Data Bit 7	0	0	Divisor Latch Access Bit (DLAB)	0	0	Received Line Signal Detect (RLSD)	Bit 7	Bit 15

* Bit 0 is the least significant bit.
It is the first bit serially received or transmitted.

Table 6.4. A summary of the register functions accessible through the divisor latch access bit (DLAB) and the register addresses. It will be helpful to relate this table to the block diagram of Figure 6.2.

Interrupt Identification Register			Interrupt Set and Reset Functions			
Bit 2	Bit 1	Bit 0	Prority Level	Interrupt Type	Interrupt Source	Interrupt Reset Control
0	0	1	−	None	None	
1	1	0	Highest	Receiver Line Status	Overrun Error or Parity Error or Framing Error or Break Interrupt	Reading the Line Status Register
1	0	0	Second	Received Data Available	Receiver Data Available	Reading the Receiver Buffer Register
0	1	0	Third	Transmitter Holding Register Empty	Transmitter Holding Register Empty	Reading the IIR Register (If Interrupt Source) or Writing into the Transmitter Holding Register
0	0	0	Fourth	MODEM Status	Clear to Send or Data Set Ready or Ring Indicator or Received Line Signal Detect	Reading the Modem Status Register

Table 6.6. This table defines the DLAB bit coding for the interrupt set and reset functions within the interrupt identification register (IIR). Note this is a read-only register.

UART Pin Descriptions

The function descriptions that follow are in alphabetical order by the identifying acronym. The corresponding pin numbers immediately follow, preceding a summary of the function. Note that in this book, "/" defines the complement—that is, the term is active low.

Function	Pin(s)	Description
A2,A1,A0	26-28	Signals applied to these three inputs select a UART register for reading from or writing to by the CPU during a transfer of data. Table 6.1 relates the address inputs to the register they select. The state of the divisor latch access bit (DLAB), the most significant bit of the line control register, determines the slection of certain UART registers. The DLAB (Bit 7 of address 011) must be set high to access the baud generator divisor latches. Note

that an active /ADS input is required when the select addresses are not stable during the duration of a read or write operation. When not a factor, this input line is tied permanently low.

/BAUDOUT	15	Baud Out Pin. This is the 16X clock signal provided to the transmitter section of the UART. The clock rate is equal to the main reference oscillator frequency divided by the specified divisor in the baud generator divisor latches. This clock signal may also be utilized by the receiver section by connection to the RCLK input, Pin 9.
CS0,CS1,	12-14	The chip is selected with CS0,CS1 high and /CS2 low.
/CS2		This enables communication between the UART and the CPU. The positive edge of an active address strobe signal latches the decoded chip select signals to complete chip selection. If /ADS is always low, the chip selects should stabilize according to the tCSW schedule (see Figure 6.7).
CSOUT	24	Chip Select Out. When this line is high, it indicates the chip has been selected by active CS0,CS1,/CS2 inputs. This line must be at logic one to enable a data transfer. The line goes low with the chip deselected.
/CTS	36	Clear-to-Send. When this pin is low, it indicates that the modem or data set is ready to exchange data. This signal is a modem status input whose condition can be tested by the CPU reading Bit 4 (CTS) of the modem status register. Bit 4 is the complement of the /CTS signal. Bit 0 (DCTS) of the modem status register indicates whether the /CTS input has changed status since the previous reading of the modem status register. /CTS has no effect on the transmitter. Note that whenever the CTS bit of the modem status register changes state, an interrupt is generated if the modem status interrupt is enabled.
D7-D0	1-8	The data bus is comprised of eight TRI-STATE[3] input/output lines. The bus provides bidirectional communications between the UART and the CPU. Data, control words, and status information are transferred via this bus.
/DCD	38	Data Carrier Detect pin. When this pin is low, it indicates the data carrier has been detected by the modem or data set. The /DCD signal is a modem status input whose condition can be tested by the CPU reading Bit 7 (DCD) of the modem status register. Bit 7 is the complement of the /DCD signal. Bit 3 (DDCD) of the modem status register indicates whether the /DCD input has changed state since the previous reading of the modem status register. /DCD has no effect on the receiver. Note that whenever the DCD bit of the modem status register changes state, an interrupt is generated if the modem status interrupt is enabled.

DDIS   23   Driver Disable. This line goes low whenever the CPU is *reading* data from the UART. It can be used to disable or control the direction of a data bus transceiver between the UART and the CPU (refer to *Figure 6.10*).

/DSR   37   Data Set Ready. When low, this indicates the modem or data set is ready to establish the communication link with the UART. This signal is a modem status input whose condition can be tested by the CPU reading Bit 5 (DSR) of the modem status register. Bit 5 is the complement of the /DSR signal. Bit 1 (DDSR) of the modem status register indicates whether the /DSR input has changed state since previous reading of the modem status register. Note that whenever the DSR bit of the modem status register changes state, an interrupt is generated if the modem status interrupt is enabled.

/DTR   33   Data Terminal Ready. When low, this indicates the modem or data set is ready to establish the communication link with the UART. The /DTR signal can be set to an active low by programming Bit 0 (DTR) of the modem control register to a high level. A master reset operation sets this signal to its inactive (high) state. Loop-mode operation holds the signal in its inactive state.

INTR   30   Interrupt. This line goes high whenever any one of the following interrupt types has an active high condition and is enabled via the IER: receiver line status, received data available, transmitter-holding register empty, and modem status. The INTR line is reset low upon the appropriate interrupt service or a master reset operation. The INS8250 and INS8250-B response to multiple interrupts differs from the INS8250A and NS16450. This is described later under the section heading "INS8250 and INS8250-B Functional Considerations."

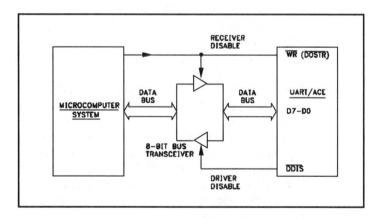

Figure 6.10. Application of DDIS to disable communications between the data bus and the UART.

MR	35	Master Reset. With this input high, all registers are cleared other than the receiver buffer, transmitter-holding register, and the divisor latches. All UART control logic is cleared. The states of outputs SOUT, INTR, /OUT1, /OUT2, /RTS, and /DTR are as shown in Table 6.2. This input is buffered with a TTL-compatible Schmitt trigger with typical 0.5V hysteresis.
/OUT1	34	Output 1. This is a user-designated output that is set to an active low by programming Bit 2 (OUT 1) of the modem control register to a logic 1 level. A master reset sets this signal to its inactive (high) state. Loop-mode operation holds it in the inactive state. In the XMOS parts, this will achieve TTL logic levels.
/OUT2	31	Output 2. This is a user-designated output that is set to an active low by programming Bit 3 (OUT 2) of the modem control register to a logic 1 level. A master reset sets this signal to its inactive (high) state. Loop-mode operation holds it in the inactive state. In the XMOS parts, this will achieve TTL logic levels.
RCLK	9	Receiver Clock. This input is the 16 X baud-rate clock for the receiver portion of the chip.
RD,/RD	22,21	Read. With RD high, or /RD low, with the chip selected, the CPU can read status information or data from the selected UART register. Note that only an active RD or /RD is required to transfer data during a read operation. Therefore, tie either the RD line permanently low or the /RD permanently high, when not being used. The early 8250-B designations for RD,/RD are DISTR, /DISTR. There is no functional difference.
/RI	39	Ring Indicator. When low, this input indicates that a telephone ringing signal has been received by the modem or data set. The /RI signal is a modem status input whose condition can be tested by the CPU reading Bit 6 (RI) of the modem status register. Bit 6 is the complement of the /RI signal. Bit 2 (TERI) of the modem status register indicates whether the /RI input signal has changed from a low to a high state since the previous reading of the modem status register. Note that whenever the RI bit of the modem status register changes from a high to a low state, an interrupt is generated if the modem status interrupt is enabled.
/RTS	32	Request-to-Send. When low, this output line informs the modem or data set that the UART is ready to exchange data. The /RTS signal can be set to an active low by programming Bit 1 (RTS) of the modem control register to a high level. A master reset operation sets this signal to its inactive (high) state. Loop mode operation holds the signal in its inactive state.

SIN	10	Serial Input. Serial data input from the communication link: modem, data set, or peripheral device.
SOUT	11	Serial Out. The composite serial data output to the communication link: modem, data set, or peripheral device. The SOUT signal is set to the marking (logic 1) state upon a master reset or when the transmitter is idle.
Vcc	40	+5V dc power supply.
Vss	20	Ground (0V) reference.
WR,/WR	19,18	Write. With WR high, or /WR low, with the chip selected, the CPU can write control words or data into the selected UART register. Note that only an active WR or /WR is required to transfer data to the UART during a write operation. Therefore, tie either the WR line permanently low or the /WR permanently high, when not being used. The early 8250-B designations for WR,/WR are DOSTR,/DOSTR. There is no functional difference.
XIN	16	External Crystal Input. This signal input is used in conjunction with XOUT to form a feedback circuit for the baud-rate generator's oscillator. If a clock signal will be generated off-chip, it should drive the baud rate generator through this pin.
XOUT	17	External Crystal Output. This signal input is used in conjunction with XIN to form a feedback circuit for the baud-rate generator's oscillator. If a clock signal will be generated off-chip, this pin is unused.

UART Registers

Table 6.4 presents a summary of the UART registers. Register access and their content is identical for all versions of the device. These are all accessible to the CPU via the eight-bit data bus. Control of the device operation is exercised via these register functions. If you find this table confusing, you're not alone.

Return to the register addresses given in Table 6.1. Let's look at the leftmost column with the heading "DLAB". Now note that addresses 000 and 001 appear twice—once at the beginning with DLAB shown as "0" and once again at the bottom with DLAB shown as "1."

When we get to the bench-top procedures this will be clearer, but what is needed—whether here on the bench or with a computer program—is to go to address 011, the line control register, and set Bit 7 to "1" with all other bits at logic "0." This puts us in the DLAB.

Now we can set the address to 000 and enter the divisor latch least significant bits; then proceed to address 001 to enter the most significant bits. With that accomplished, we can return to address 011 and set Bits 0 through 6 for our data communications with Bit 7 now set to 0. With this in mind, we can investigate the register functions. Note that resetting the device will not affect the divisor latch settings. (This is true only for the XMOS devices; the CMOS require reentering the divisor-latch values.)

Line Control Register

The line control register, address 011, specifies the format of the asynchronous data-communication exchange. Table 6.5 defines the coding of Bits 0 and 1 of this register for selection of the required character length in the exchange.

Bit 2 specifies the number of stop bits transmitted and received in each serial character. If Bit 2 is a logic 0, then one stop bit is generated or checked in the transmitted data. If Bit 2 is a logic 1 when a five-bit word length is selected via bits 0 and 1, then one and one-half stop bits are generated. If bit 2 is a logic 1 whenever one of the other three length options is selected, two stop bits are selected. Note that the receiver checks only the first stop bit, regardless of the number of stop bits selected.

Bit 3 is the parity enable bit. When this bit is a logic 1, a parity bit is transmitted and tested for in the received data between the last data word bit and the stop bit. The parity bit is used to produce either an even or odd number of 1s when the data word bits and the parity bit are summed.

Bit 4 is the even parity select bit. When bit 3 is a logic 1 and bit 4 is a logic 0, an odd number of logic 1s is transmitted or checked for. When bit 3 is a logic 1 and bit 4 is a logic 1, an even number of logic 1s is transmitted or checked for.

Bit 5 is the stick parity bit. When bits 3, 4, and 5 are logic 1, the parity bit is transmitted and checked as a logic 0. If bits 3 and 5 are 1 and bit 4 is a logic 0, then the parity is transmitted and checked as a logic 1. If bit 5 is a logic 0, then the stick parity is disabled.

Bit 6 is the break control bit. It causes a break condition to be transmitted by the UART. When it is set to a logic 1, the serial output (SOUT) is forced to the spacing (logic 0) state. The break is disabled by clearing Bit 6 to a logic 0. The break control bit acts only on SOUT and has no effect on the transmitter logic.

(Note: This feature enables the CPU to alert a terminal in a computer communications system. If the following sequence is used, no erroneous or extraneous characters will be transmitted because of the break.

1. Load in all 0s, pad character, in response to THRE.

2. Set break after the next THRE.

3. Wait for the transmitter to be idle (TEMT=1), and clear break when normal transmission has to be restored.

During the break, the transmitter can be used as a character timer to accurately determine the break duration.)

Bit 7 is the divisor-latch access bit (DLAB). It must be set high (logic 1) to access the divisor latches of the baud generator during a read or write operation. It must be set low (logic 0) to access the receiver buffer, the transmitter-holding register or the interrupt enable register and the line and modem status or control registers.

Programmable Baud Generator

The UART contains a programmable baud-rate generator capable of taking any clock input from DC to 3.1 MHz and dividing it by any divisor from 1 to 2e16-1. Clock circuits are depicted in Figure 6.5. Table 6.3a and 6.3b define available baud rates with a 1.8432-MHz and 3.072-MHz crystal, respectively.

The output frequency of the baud-rate generator is 16 X the baud [divisor # = (frequency input) ÷ (baud rate X 16)]. Two eight-bit latches store the divisor in a 16-bit binary format. These divisor latches must be loaded during initialization in order to ensure proper operation of the baud generator. Upon loading either of the divisor latches, a 16-bit baud counter is immediately loaded.

(Note: The maximum operating frequency of the baud generator is 3.1 MHz. When using divisors of 3 and below, however, the maximum frequency is equal to the divisor in megahertz. For example, if the divisor is 1, then the maximum frequency is 1 MHz. In no case should the data rate be greater than 56K baud.)

Line Status Register

This eight-bit register provides status information to the CPU concerning the data transfer. Table 6.4 shows the content of each register bit as follows.

Bit 0 is the receiver data-ready (DR) indicator. This bit is set to a logic 1 whenever a complete incoming character has been received and transferred into the receiver buffer register. This bit is reset to a logic 0 by reading the data in the receiver buffer register.

Bit 1 is the overrun error (OE) indicator. This bit indicates that data in the receiver buffer register was not read by the CPU before the next character was transferred into the receiver buffer register, thereby destroying the previous character. The OE indicator is set to a logic 1 upon detection of an overrun condition and is reset to a logic 0 whenever the CPU reads the content of the line status register.

Bit 2 is the parity error (PE) indicator. This bit indicates that the received data character does not have the correct even or odd parity as selected by the even-parity select bit. The PE is set to a logic 1 upon detection of a parity error and is reset to a logic 0 whenever the CPU reads the content of the line status register.

Bit 3 is the framing error (FE) indicator. This bit indicates that the received data did not have a valid stop bit. This bit is set to a logic 1 whenever the stop bit following the last data bit or parity bit is a logic 0 (spacing level).

The FE indicator is reset to a logic 0 whenever the CPU reads the content of the line status register. The UART will try to resynchronize after a framing error. To do this, it assumes that the framing error was due to the next start bit, so it samples this "start" bit twice and then takes in the "data."

Bit 4 is the break interrupt (BI) indicator. This bit is set to a logic 1 whenever the received data input is held in the spacing (logic 0) state for longer than a full word transmission time—that is, the total time of start bit + data bits + parity + stop bits.

The BI indicator is reset to a logic 0 whenever the CPU reads the content of the line status register. Restarting after a break is received requires the SIN pin to be logical 1 for at least ½ bit time.

(Note: Bits 1 through 4 are the error conditions that produce a receiver-line status interrupt whenever any of the corresponding conditions are detected and the interrupt is enabled.)

Bit 6 is the transmitter empty (TEMT) indicator. This bit is set to a logic 1 whenever the transmitter holding register (THR) and the transmitter shift register (TSR) are both empty. It is reset to logic 0 whenever either the THR or the TSR contains a data character.

Bit 7 is permanently set to logic 0.

(Note: The line status register is intended for read operations only. Writing to this register is not recommended, as this operation is only used for factory testing.)

Interrupt Identification Register

In order to minimize software overhead during data transfers, the UART prioritizes interrupts into four levels and records these in the interrupt identification register. The four levels of interrupt conditions are—in order of priorities—receiver line status, received data ready, transmitter holding register empty, and modem status.

When the CPU accesses the IIR, the UART freezes all interrupts and indicates the highest-priority pending interrupt to the CPU. While the CPU access is occurring, the UART records new interrupts but does not change its current indication until the access is complete. Table 6.5 shows the contents of the IIR. Details on each bit follow.

• Bit 0 can be used in an interrupt environment to indicate whether an interrupt condition is pending. When this bit is a logic 0, an interrupt is pending and the IIR contents may be used as a pointer to the appropriate interrupt service routine. When this bit is a logic 1, no interrupt is pending.

• Bits 1 and 2 are used to identify the highest-priority interrupt pending, as indicated in Table 6.6.

• Bits 3 through 7 are always at logic 0.

Interrupt Enable Register

This register enables the four types of UART interrupts. Each interrupt can individually activate the interrupt (INTR) output signal. It is possible to totally disable the interrupt system by resetting Bits 0 through 3 of the interrupt enable register (IER).

Similarly, setting bits of this register to a logic 1 enables the selected interrupt(s). Disabling an interrupt prevents it from being indicated as active in the IIR and from activating the INTR output signal. All other system functions operate in their normal manner, including the setting of the line status and modem status registers. Table 6.5 shows the contents of the IER. Details of each bit follow.

• Bit 0 enables the received-data available interrupt when set to logic 1.

• Bit 1 enables the transmitter-holding register empty interrupt when set to logic 1.

- Bit 2 enables the receiver line status onterrupt when set to logic 1.

- Bit 3 enables the modem status interrupt when set to logic 1.

- Bits 4 through 7 are always logic 0.

Modem Control Register

This register controls the interface with the modem or data set (or a peripheral device emulating a modem). The contents of the modem control register (MCR) are indicated in Table 6.5. Details on each bit follow.

Bit 0 controls the data terminal ready (/DTR) output. When this bit is set to a logic 1, the /DTR output is forced to a logic 0. When this bit is reset, the /DTR output is forced to a logic 1.

(Note: The /DTR output of the UART may be applied to an EIA inverting the driver (such as the DS1488) to obtain the proper polarity input at the succeeding modem or data set.)

Bit 1 controls the request to send (/RTS) output. When this bit is set to a logic 1, the /RTS output is forced to a logic 0. When this bit is reset, the /RTS output is forced to a logic 1.

Bit 2 controls the Output 1 (/OUT1) signal, which is an auxiliary user-designated output. When this bit is set to a logic 1, the /OUT1 output is forced to a logic 0. When this bit is reset, the /OUT1 output is forced to a logic 1.

Bit 3 controls the Output 2 (/OUT2) signal, which is an auxiliary user-designated output. When this bit is set to a logic 1, the /OUT2 output is forced to a logic 0. When this bit is reset, the /OUT2 output is forced to a logic 1.

Bit 4 provides a local feedback feature for diagnostic testing of the UART. When this bit is set to logic 1 the following occur.

- The transmitter serial output (SOUT) is set to the marking (logic 1) state.

- The receiver serial input (SIN) is disconnected.

• The output of the transmitter shift register is "looped back" into the receiver shift register input.

• The four modem control inputs (/DSR, /CTS, /RI, /DCD) are disconnected.

• The four modem control outputs (/DTR, /RTS, /OUT1, /OUT2) are internally connected to the four modem control inputs.

The modem control output pins are forced to their inactive state (high). In the diagnostic mode, data that is transmitted is immediately received. This feature allows the processor to verify the transmit-and-received-data paths of the UART.

In the diagnostic mode, the transmitter and receiver interrupts are fully operational. The modem control interrupts are also operational, but the interrupts sources are now the lower four bits of the modem control register, instead of the four modem control inputs. The interrupts are still controlled by the interrupt enable register.

Bits 5 through 7 are permanently set to logic 0.

Modem Status Register

This register provides the current state of the control lines from the modem (or peripheral device) to the CPU. In addition to this current-state information, four bits of the modem status register provide change information. These bits are set to a logic 1 whenever a control input from the modem changes state. They are reset to logic 0 whenever the CPU reads the modem status register. Table 6.5 shows the status of the MSR. Details on each bit follow.

Bit 0 is the delta clear-to-send (DCTS) indicator. This bit indicates that the /CTS input to the chip has changed state since the last time it was read by the CPU.

Bit 1 is the delta data set ready (DDSR) indicator. This bit indicates that the /DSR input to the chip has changed state since the last time it was read by the CPU.

Bit 2 is the trailing-edge of ring-indicator (TERI) detector. This bit indicates that the /RI input to the chip has changed from a low to a high state.

Bit 3 is the delta data-carrier detect (DDCD) indicator. This bit indicates that the DCD input to the chip has changed state.

(Note: Whenever Bit 0, 1, 2, or 3 is set to logic 1, a modem status interrupt is generated.)

Bit 4 is the complement of the clear-to-send (/CTS) input. If Bit 4 is set to a logic 1, this bit is equivalent to RTS in the MCR.

Bit 5 is the complement of the data-set-ready (/DSR) input. If Bit 4 is set to a logic 1, this bit is equivalent to DTR in the MCR.

Bit 6 is the complement of the ring-indicator (/RI) input. If Bit 4 is set to a logic 1, this bit is equivalent to OUT1 in the MCR.

Bit 7 is the complement of the data-carrier detect (DCD) input. If Bit 4 is set to a logic 1, this bit is equivalent to OUT2 in the MCR.

Manufacturing Summary[4]

National Semiconductor manufactured four versions of the XMOS INS8250 and NS16450. Functionally, these parts appear to be the same; however, there are differences that designers and users need to be aware of[5]. In this section, the characteristics of each—with details of function and time—are discussed. A summary of the devices follows.

1. The INS8250. This is the original device manufactured. It is the same part as the INS8250-B, but with faster CPU bus timing.

2. The INS8250-B. This is the slower speed (CPU bus timing) version of the INS8250. It is used by many popular 8088-based computers.

3. The INS8250A. This device is a revision of the INS8250 using the more advanced XMOS process. The INS8250A performance is better than either of 1 and 2 above due to the redesign, which incorporated

finer process controls in their manufacture. This part was used in many 8086-based computers.

4. The NS16450. This is the faster speed (CPU bus timing) version of the INS8250A. It was used in many 80286-based computers.

In addition to the XMOS devices, there are:

1. The INS82C50A. The CMOS version of the INS8250A. It functions identically and, for most AC parameters, has the same timing application as the INS8250A. It draws approximately 1/10 (10 mA) of the maximum operating current of the INS8250A.

2. The NS16C450. The CMOS version of the NS16450. It functions identically and, for most AC parameters, has the same timing application as the INS8250A. It draws approximately 1/12 (10 mA) of the maximum operating current of the NS16450.

INS8250 and INS8250-B Functional Considerations

For these parts, it's important to be aware of the following.

1. When multiple Interrupts are pending, the interrupt line (INTR) pulses low after each interrupt, instead of remaining high continuously. This may not be a problem in normal operation, but it is a condition necessary for compatibility with some 8086- and 80286-based microcomputers that use an edge-triggered ICU.

2. Bit number 6 (TSRE) of the line status register is set as soon as the transmitter shift register empties—whether or not the transmitter holding register contains a character. Bit number 6 is then reset when the transmitter shift register is reloaded. This may not be a problem in normal operation, but it is a function tested on some 8088-based microcomputer systems diagnostic procedures.

3. In loopback mode, the modem control outputs /RTS, /DTR, /OUT1, and /OUT2 remain connected to the associated modem control register bits.

INS8250A and NS16450 Functional Considerations

For these parts, it's important to be aware of the following.

1. The loopback diagnostic function sets the modem control outputs /RTS, /DTR, /OUT1, and /OUT2 to their inactive state (logic "1") so they will not send out spurious signals.

2. A one-byte scratch-pad register is included at location 111. This register is not present on the INS8250 and INS8250-B.

3. When multiple interrupts are pending, the interrupt line remains high, rather than pulsing low after each interrupt is serviced. The INS8250A and NS16450 have level-sensitive interrupts, as opposed to edge-triggered interrupts. This requires a change in the UART driver software or associated hardware, if these devices are used with microcomputers employing edge-triggered ICUs.

4. Bit 6 of the line status register is set to logic 1 when both the transmitter-holding register and shift register are empty. This causes these devices to be incompatible with some software using this bit.

INS8250A and NS16450 Timing Considerations

For these parts, it's important to be aware of the following.

1. A start bit will be sent at typically 16 clocks (1 bit time) after the WRTHR signal goes active.

2. The leading edge of WRTHR resets THRE and TEMT.
3. All of the line-status errors and the received data flag (DR, data ready) are set during the time of the first stop bit.

4. TEMT is set two RCLK clock periods after the stop bit(s) are sent.

5. The modem control register updates the modem outputs on the trailing edge of the WRMCR.

All of the preceding is applicable to the corresponding CMOS parts. In addition, the following items specify differences between XMOS and CMOS parts and are applicable to the CMOS only.

1. Anytime a reset pulse is issued to the INS82C50A or NS16C450, the divisor latches must be rewritten with the appropriate divisors in order to start the baud-rate generator.

2. tSI is from 16 to 48 RCLK cycles in length.

Software Compatibility

Existing computers using the INS8250-B are required to use two of the foregoing considerations. These detect multiple pending interrupts and test the baud rate. These concerns were eliminated in the revision part and all succeeding parts.

Thus, a software or hardware change is required for the utilization of the more recent parts as replacements for the INS8250-B. If the target system services the UART via polling rather than interrupts, the later parts will be plug-in replacements.

Using the INS8250A, NS16450, INS82C50A, and NS16C450 With Edge-Triggered ICUs

Using these devices with an edge-triggered ICU—as with some micro-computers—requires a signal edge on the INTR pin for each pending UART interrupt.

Otherwise, when multiple interrupts are pending, the interrupt line will be constantly high active and the edge-triggered ICU will not request additional services for the interrupt.

UART CHIPS IN PC OR AT SYSTEMS	
CHIP	DESCRIPTION
8250	IBM used this chip in the PC serial port card. The chip has several bugs, none of which are serious. The PC and XT ROM BIOS are written to anticipate at least one of the bugs. This chip was replaced by the 8250B.
8250A	This upgraded chip fixes the several bugs in the 8250, including one in the interrupt enable register, but because the PC and XT ROM BIOS expect the bug this chip does not work properly with those systems. The 8250A should work properly in an AT system that does not expect the bug, but does not work adequately at 9600 bps.
8250B	The last version of the 8250 fixes bugs from the previous two versions. The interrupt enable bug in the original 8250, expected by the PC and XT ROM BIOS software, has been put back into this chip, making the 8250B the most desirable chip for any non-AT serial port application. The 8250B chip may work in an AT under DOS, but does not run properly at 9600 bps.
16450	IBM selected the higher-speed version of the 8250 for the AT. Because this chip has fixed the interrupt enable bug, the 16450 does not operate properly in many PC or XT systems, because they expect the bug to be present. OS/2 requires this chip as a minimum, or the serial ports do not function properly. It also adds a scratch pad register as the highest register. The 16450 is used primarily in AT systems because of its increase in throughput over the 8250B.
16550	This newer UART improves on the 16450. This chip cannot be used in a FIFO buffering mode because of problems with the design, but it does enable a programmer to use multiple DMA channels and thus increase throughput on an AT or or higher class computer system.
16550A	This chip is a faster 16450 with a built-in 16-character Transmit and Receive FIFO buffer that works. It also allows multiple DMA channel access. This is the chip for use in communication systems operating at 9600 bps or higher. It can greatly increase communications speed and eliminates lost characters and data at higher speeds.

Table 6.7. This table provides additional information on the functional concerns of INS8250/NS16450 series of UARTS.

HOW TO USE THE DATA ENTRY R/W SWITCH WITH THE 8250/16450

With this IC data entry on the bus is not related to the clock.

1. To write to the Data Bus:
   Ensure the Data Entry R/$\overline{W}$ switch is in the R position.
   Establish the Data Pattern on the Data Entry Module
2. Use the Address Select DIP switch to enter the required address.
3. Toggle the Data Entry R/$\overline{W}$ switch to the $\overline{W}$ position.
   Observe the pattern appears on LED Module 4
4. Toggle the 8250 Write switch.
5. Return the Data Entry R/$\overline{W}$ switch to the R position.
   Unless noted otherwise a successful write will be followed by a display of the entered data with the switch in the R position.

Table 6.8. Instructions on usage of the data-entry and R/W toggle switches found on the bus-entry module.

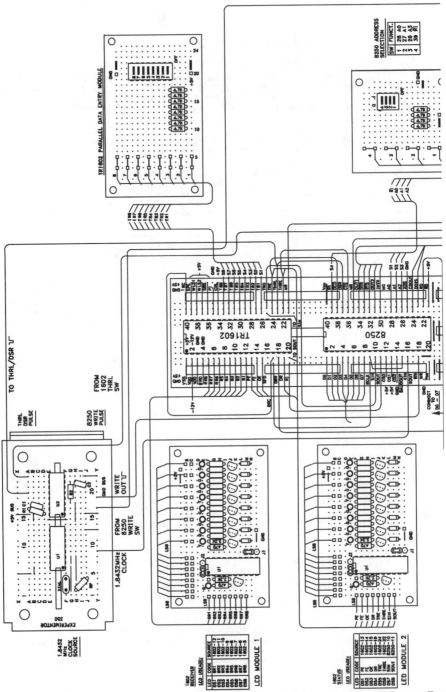

Figure 6.13. *(Continued on next page.)* **The author's bench-top module layout and and wiring configuration for operation of the 8250/16450 UART.**

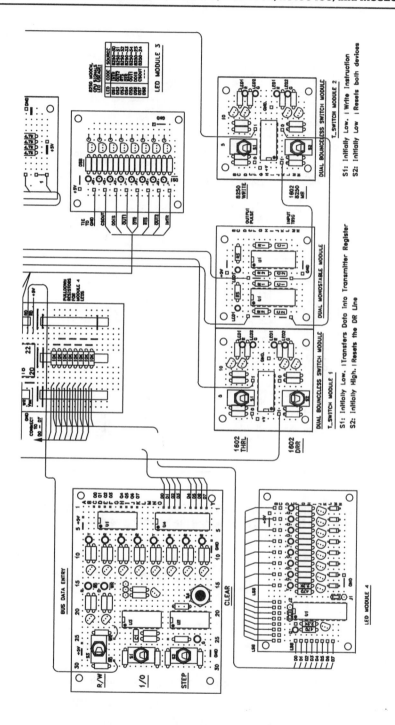

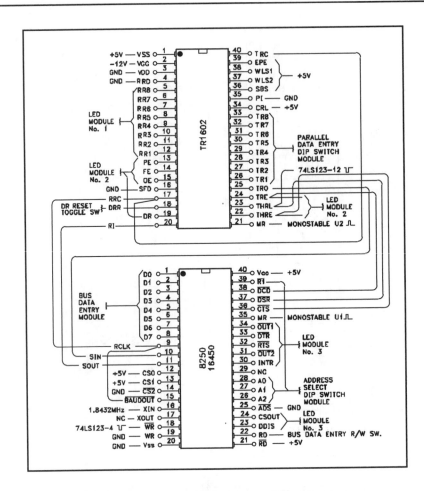

Figure 6.12. An interconnecting planning diagram for linking the TR1602 with the 8250/16450. You should use this diagram along with Figure 6.13 when wiring the operating configuration.

Creating an Interrupt Edge Via Software

This is done by disabling and then reenabling UART interrupts via the interrupt enable register (IER) before a specific UART interrupt-handling routine (line-status errors, received data available, transmitter holding register empty, or modem status) is exited. To disable interrupts, write H'00 to the IER. To reenable interrupts, write a byte containing "1's" to the IER bit positions whose interrupts are supposed to be enabled.

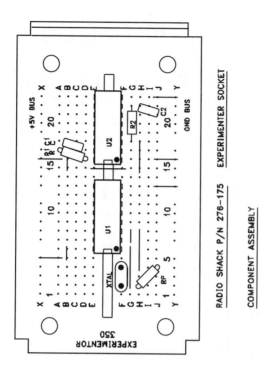

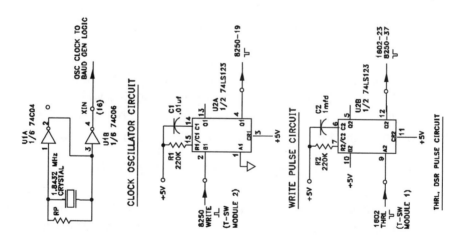

Figure 6.11. Circuitry and prototype board layout for the 1.8432-MHz clock and the dual monostable for the 8250/16450 data bus write (/WR) and 1602 transmitter-holding function load (THRL) functions.

Creating an Interrupt Edge in Hardware

This is done externally to the UART. One approach is to connect the INTR pin of the UART to the input of AND gate. The other input of this AND gate is connected to a signal that will always go low active when the UART is accessed. The gate output is then used as the interrupt to the ICU.

(Note: This simple hardware recommendation will result in one invalid interrupt being generated, so the software routine must be able to handle this. This approach was tested in a few 8088-based computer systems.)

In addition to the considerations described, *Table 6.7* describes further concerns to the users of these devices.

Bench-Top Operation of the INS8250-B/NS16450 UART

Figure 6.13 is my bench test setup for these devices. You will notice that the TR1602 is a part of this operation. The two work together very well. The planning diagram of *Figure 6.12* will be helpful in the wiring of your own setup.

Previous chapters provided suggestions for the construction of the modules and their corrugated cardboard mountings, so that will not be repeated here. Do take care with the connections, double-checking as you go, as with the congestion it is easy to poke a wire into an incorrect point.

Review the switch usage described in *Table 6.8*, as there is a departure from previous practice with these devices. Entering write commands is not linked to the clock here on the bench. Also, I found that these wouldn't respond to the high-to-low transition of the bounceless switches employed with the modules. For that reason, you will find a monostable circuit for this purpose, as shown in *Figure 6.11*. The sharp negative going pulse of the 74LS123 does the job very nicely. The other half of the mono loads the TR1602 transmitter-holding register (THRL).

INITIALIZING THE 8250/16450 UART

CAUTION: Review Table 6.7 before beginning this procedure
to ensure correct R/W switch and DIP switch useage.

NOTE: The LEDs in Modules 1 and 2 may show random
displays on power up.

1. Access the Line Control Register and ☐
   enable DLAB for setting the Baud Rate
   Generator Divisor Latches.
   The required Baud Rate Divisor is Decimal 7200
   Binary 0001 1100 0010 0000

```
       OFF
1 A0 [ ⬛ ]  1        | 7 6 5 4 3 2 1 0 | OPERATION
2 A1 [ ⬛ ]  1        | 1 0 0 0 0 0 0 0 | Access the DLAB
3 A2 [ ⬛ ]  0
4 -- [ ⬛ ]  -    ☐   BUS DATA ENTRY = 1000 0000 ☐
       1  0
   REG ADDRESS
   DIP SW SETTINGS
```

2. Set the Lower Baud Rate Generator Latch
   to the Least Significant Baud Rate Divsior   ☐

```
       OFF
1 A0 [ ⬛ ]  0        | 7 6 5 4 3 2 1 0 | OPERATION
2 A1 [ ⬛ ]  0        | 0 0 1 0 0 0 0 0 | LSB Entry
3 A2 [ ⬛ ]  0
4 -- [ ⬛ ]  -    ☐   BUS DATA ENTRY = 00100 0000 ☐
       1  0
   REG ADDRESS
   DIP SW SETTINGS
```

3. Set the Upper Baud Rate Generator Latch
   to the Most Significant Baud Rate Divisor   ☐

```
       OFF
1 A0 [ ⬛ ]  1        | 7 6 5 4 3 2 1 0 | OPERATION
2 A1 [ ⬛ ]  0        | 0 0 0 1 1 1 0 0 | MSB Entry
3 A2 [ ⬛ ]  0
4 -- [ ⬛ ]  -    ☐   BUS DATA ENTRY = 0001 1100 ☐
       1  0                Observe a pulse train at pin 15 (BAUDOUT)
   REG ADDRESS              of 256 Hz ☐
   DIP SW SETTINGS         LED Module 2, DS8 Blinking (INS8250-B only)
```

4. Access the Line Control Register to format
   the Asynchronous Data Communications protocol. ☐

```
       OFF
1 A0 [ ⬛ ]  1        | 7 6 5 4 3 2 1 0 | OPERATION
2 A1 [ ⬛ ]  1        | 0 0 1 1 1 0 1 1 | Data Communications
3 A2 [ ⬛ ]  0
4 -- [ ⬛ ]  -    ☐   BUS DATA ENTRY = 0011 1011 ☐
       1  0                Bit 5: Stick Parity
   REG ADDRESS             Bit 4: Select Even Parity
   DIP SW SETTINGS         Bit 3: Enable Parity
                          Bits 1,0: Word Length = 8 Bits
```

5. Access the Interrupt Enable Register. ☐

```
       OFF
1 A0 [ ⬛ ]  1        | 7 6 5 4 3 2 1 0 | OPERATION
2 A1 [ ⬛ ]  0        | 0 0 0 0 1 1 1 1 | Interrupts enabled
3 A2 [ ⬛ ]  0
4 -- [ ⬛ ]  -    ☐   BUS DATA ENTRY = 0000 1111 ☐
       1  0
   REG ADDRESS
   DIP SW SETTINGS
```

8250/16450
INITIAL STATUS OF LED MODULE 3

DS1 INTR – OFF
DS2 OUT1 – ON
DS3 DTR   – ON
DS4 RTS   – ON
DS5 OUT2 – ON
DS6 DDIS – OFF In Read
DS6 DDIS – ON In Write
DS7 CSOUT – ON

1602/8250
INITIAL STATUS OF LED MODULE 2

DS1 PE   – OFF
DS2 FE   – OFF
DS3 OE   – OFF
DS4 DR   – OFF
DS5 TRE – ON
DS6 THRE – ON
DS7 SIN – ON (8250)
DS8 SOUT – ON* (8250)
* 8250-B may slow blink

1602
INITIAL STATUS OF LED MODULE 1

All dark or random

6. Proceed to the Loop or
   Run Procedure.

Figure 6.14. The initialization procedure for the INS8250-B and NS16450/INS8250A UARTs.

8250/16450 COMMUNICATION PROCEDURE WITH THE TR1602 UART

CAUTION: Review Table 6.7 before beginning this procedure to ensure correct R/W switch and DIP switch useage.
The Initialization procedure must have been performed.

1. Setup the TR1602 for Data Communications.

OFF
1 TR1 1
2 TR2 0
3 TR3 1
4 TR4 1
5 TR5 1
6 TR6 0
7 TR7 0
8 TR8 1
1 0

Optional Data Entry pattern for the TR1602 Parallel Data DIP switch ☐

2. Enable the Data Communications via the MODEM Control Register. ☐

OFF
1 A0 0
2 A1 0
3 A2 1
4 -- – ☐
1 0
REG ADDRESS
DIP SW SETTINGS

7 6 5 4 3 2 1 0	OPERATION
0 0 0 0 1 1 1 1	Normal operation mode

BUS DATA ENTRY = 0000 1111 ☐

Observe status of LED Module 3 indicators:
OUT1 – OFF
DTR – OFF
RTS – OFF
OUT2 – OFF
INTR – OFF (INS8250-B)
INTR – ON (NS16450) ☐

3. Enter the Data for Transmission on the Bus. ☐

OFF
1 A0 0
2 A1 0
3 A2 0
4 -- – ☐
1 0
REG ADDRESS
DIP SW SETTINGS

7 6 5 4 3 2 1 0	OPERATION
1 1 0 1 1 0 1 1	Optional Pattern

BUS DATA ENTRY = 1101 1011 ☐

LED Module 2, DSB blinks with Data ☐
LED Module 1 LEDs display Bus Data ☐
Toggle DRR ☐
Toggle THRL ☐
LED Module 2, DS7 blinks with Data ☐
The READ display now matches the TR1602 Data ☐
The Interrupt LED is not on ☐
If 8250 the INTR LED is dark ☐
If 16450 the INTR LED is On ☐

Continue with these procedures with the INS8250-B

4. Read the MODEM Status Register. ☐

OFF
1 A0 0
2 A1 1
3 A2 1
4 -- – ☐
1 0
REG ADDRESS
DIP SW SETTINGS

7 6 5 4 3 2 1 0	OPERATION
0 0 0 0 0 0 0 0	MODEM Status

BUS DATA ENTRY = Read Only ☐

5. Read the Line Status Register. ☐

OFF
1 A0 1
2 A1 0
3 A2 1
4 -- – ☐
1 0
REG ADDRESS
DIP SW SETTINGS

7 6 5 4 3 2 1 0	OPERATION
0 0 0 0 0 0 0 0	Line Status

BUS DATA ENTRY = Read Only ☐

Continue with these procedures with the INS8250A or NS16450

4. Read the MODEM Status Register. ☐

OFF
1 A0 0
2 A1 1
3 A2 1
4 -- – ☐
1 0
REG ADDRESS
DIP SW SETTINGS

7 6 5 4 3 2 1 0	OPERATION
0 0 0 0 1 0 1 1	MODEM Status

BUS DATA ENTRY = Read Only ☐

5. Read the Line Status Register. ☐

OFF
1 A0 1
2 A1 0
3 A2 1
4 -- – ☐
1 0
REG ADDRESS
DIP SW SETTINGS

7 6 5 4 3 2 1 0	OPERATION
0 1 1 0 0 0 0 0	Line Status

BUS DATA ENTRY = Read Only ☐
Bit 5: THRE
Bit 6: TEMT
Bit 1, 2, or 3 On is an error indication

Figure 6.15. The run procedure for the INS8250-B and NS16450/INS8250A UARTs.

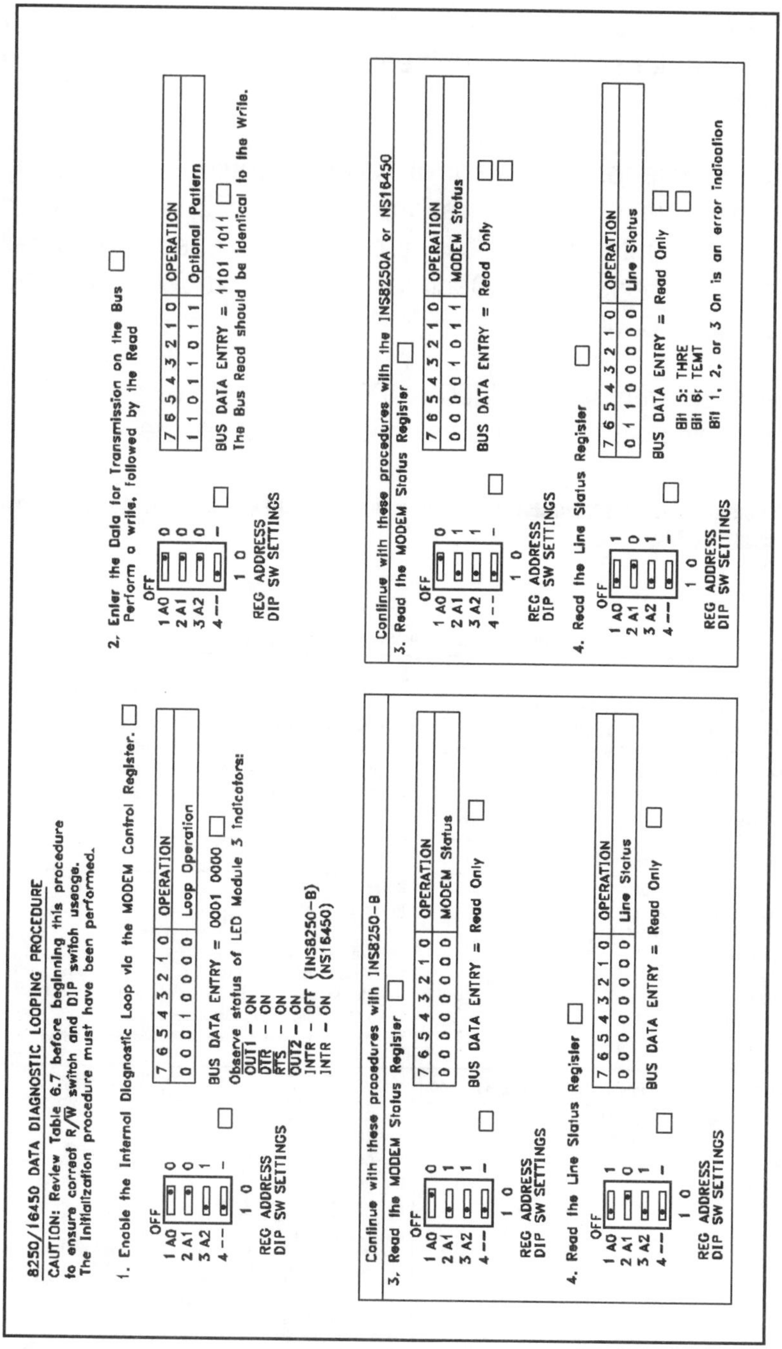

Figure 6.16. The loop procedure for the INS8250-B and NS16450/INS8250A UARTs.

The 1.8432-MHz clock circuitry is also shown with this drawing, as is the layout on an experimenter socket. It didn't seem worthwhile to make a permanent module of these.

I chose a baud-generator divisor of 7200 to provide a sufficiently low frequency to observe transfer operations with the LED indicators for SIN and SOUT. Individual bits will still go by too quickly to observe in detail, but the operation will be observable. Decimal 7200 is hexadecimal 1C20 = binary 0001 1100 0010 0000.

The conversion arithmetic is 7200 = 4096 + 2048 + 1024 + 32. The /BAUD-OUT frequency is 256 Hz, which is divided by 16 internally to yield 16 Hz. If

Table 6.9. A summary of control register functions: the interrupt enable (IER), the line control (LCR), and the modem control (MCR).

Bit	Register/Function	Operation
	CONTROL REGISTER FUNCTION SUMMARY	
	Interrupt Enable Register IER Address: 001	
0	Received Data Available Interrupt	Interrupt Enabled when set to logic 1
1	Transmitter Holding Register Empty Interrupt	Interrupt Enabled when set to logic 1
2	Receiver Line Status Interrupt	Interrupt Enabled when set to logic 1
3	MODEM Status Interrupt	Interrupt Enabled when set to logic 1
4-7	Always 0	
	Line Control Register LCR Address 011	
0	Word Length Select	Refer to Table 6.6
1	Word Length Select	Refer to Table 6.6
2	Specifies number of Stop Bits	0=1, 1=1 1/2 if 5bit data, else 2
3	Parity Enable	1=Parity bit generated
4	Even Parity Select	Bits 3 and 4=1 even parity, 1 and 0, odd parity
5	Stick Parity	1 with Bit 3=1 parity detected as 0 if Bit 4=1
6	Set Break	1=SOUT forced Low, 0=Disabled
7	DLAB Access	1= Access to Baud Rate Gen Divisor Latches
	MODEM Control Register MCR Address 100	
0	Data Terminal Ready (DTR)	1= DTR forced LOW
1	Request To Send (RTS)	1= RTS forced LOW
2	OUT 1	1= OUT1 forced LOW
3	OUT2	1= OUT2 forced LOW
4	LOOP	1= following procedure: SOUT set to logic 1 SIN disconnected Transmitter Shift Register output looped back into Receiver Shift Register input. MODEM control inputs CTS, DSR, DCD, RI disconnected MODEM control outputs DTR, RTS, OUT1, OUT2 internally connected to the four MODEM control inputs. Transmitted data is immediately received. Receiver and Transmitter Interrupts are operational. MODEM control Interrupts are operational, sources are the four lower bits of the MCR. Interrupts still controlled by the IER.
5-7	Always 0	

	STATUS REGISTER FUNCTION SUMMARY	
Bit	Register/Function	Operation
	Interrupt Identification Register IIR Address 010	
0	Interrupt Status	0 if Interrupt pending
1	Interrupt Identification, bit 0	Establishes priority with bit 2
2	Interrupt Identification, bit 1	Establishes priority with bit 1
3–7	Always 0	
	Line Status Register LSR Address 101	
0	Data Ready DR	1= complete incoming char received and is in the Receiver Buffer Register Bit is reset by CPU read or reset input
1	Overrun Error OE	1= Rcvr Buf Data overrun with new data
2	Parity Error PE	1= Received character has incorrect parity
3	Framing Error FE	1= Received character has incorrect stop bit
4	Break Interrupt BI	1= Received character held in 0 state beyond the full word transmission time
5	Transmitter Holding Register Empty THRE	1= on transfer to the Shift Reg, issues interrupt
6	Transmitter Empty TEMT	1= Transmitter Shift Register is idle. Is reset on data transfer from the Holding Register
7	Always 0	
	MODEM Status Register MSR Address 110	
0	Data Clear To Send (DCTS)	1= $\overline{CTS}$ input has changed state since CPU read
1	Delta Data Set Ready (DDSR)	1= $\overline{DSR}$ input has changed state since CPU read
2	Trailing Edge Ring Indicator (TERI)	1= indicates RI change, logic 1 to 0
3	Delta Data Carrier Detect (DDCD)	1= $\overline{DCD}$ input has changed state
		NOTE: Whenever an above bit is set, a MODEM Status Interrupt is generated
4	Clear To Send (CTS)	Complements the $\overline{CTS}$ input
5	Data Set Ready (DSR)	Complements the $\overline{DSR}$ input
6	Ring Indicator (RI)	Complements the $\overline{RI}$ input
7	Delta Carrier Detect (DCD)	Complements the $\overline{DCD}$ input

Table 6.10. A summary of status register functions: the interrupt identification (IER), the line status (LSR), and the modem status (MSR).

you have an oscilloscope or frequency counter, they are helpful in assuring the correct frequency, but not essential. If you're careful in entering the data on the bus, it will come out correctly.

There are three test procedures for these devices detailed in *Figures 6.14, 6.15,* and *6.16. Table 6.9* provides a summary of control register functions for the interrupt enable (IER), the line control (LCR), and the modem control (MCR). *Table 6.10* provides a summary of the status register functions: the interrupt identification (IER), the line status (LSR), and the modem status (MSR). You will find it convenient to refer to these summaries as you work with the three test procedures.

The first of these, Figure 6.16, provides the device-initialization routine. As mentioned previously, we access the line control register (LCR) at 011 and set Bit 7 to logic 1 for access to the divisor latches. After entering the latch values, we return to the LCR to set the protocols and exit the divisor latches. The procedure sets the data word at eight bits. You may want to experiment with other values.

The next procedure can be either of the remaining two, run or loop—whichever you desire to perform next. These can be run in either sequence without having to reinitialize. If you perform a reset, however, steps four and five of Figure 6.14 will need reentering before proceeding.

You will notice that the run and loop procedures have boxes for the INS8250-B and the NS16450 (also the INS8250A, should you prefer it). This seemed the best approach to dealing with the differences noted previously between these parts.

Should you wish to use the NS16C450 or INS82C50A, either replace the 74LS123 with a 74C221 or 74HCT123 dual monostable or include a pullup on the 74LS123 outputs. Recall that following the reset of a CMOS device the divisor latches must be reentered.

The procedures as given do not explore all the possibilities of these devices. They will reliably perform the data communication between the TR1602, but the interrupts will not be all that could be hoped for, in particular with INS8250-B. The handshaking is embedded with the TR1602 interface; if desired, you can replace these functions with another four-position DIP switch module and enter them yourself. It will be informative to change data values for repeat transfers to better see how these transactions take place. Also, I have found consistency to be a problem at times, particularly with the INS8250-B. If this is the first device you have selected for bench operation, it will take a little experience to enter the data correctly. Just be patient, do a restart, and continue.

Each procedure is designed to be self-explanatory. Even so, it will be helpful to review the text on related features and the procedures with care before and/or during your operation of the device. I have found these to be challenging devices to master. But I have also found satisfaction in the achievement. I hope you will also.

References

1. National Semiconductor Corporation, *Data Communications Local Area Networks/UARTs,* "INS8250, INS8250-B Universal Asynchronous Receiver/ Transmitter," 1990, pp. 4-3 to 4-18.

2. Ibid, "INS16450, INS8250A, INS16C450, INS82C50A Universal Asynchronous Receiver/Transmitter," pp. 4-19 to 4-35.

3. Registered trademark of National Semiconductor Corporation.

4. Ibid, Martin S. Michael, National Semiconductor Application Note 493, pp. 4-84, 85.

5. Scott Mueller, *Upgrading and Repairing PCs,* Que, 8th Edition, Table 11.4, "UART Chips in PC or AT Systems," p. 434.

6. Reference for Figure 6.1: National Semiconductor Corporation, *Data Communications Local Area Networks/UARTS,* "INS8250, INS8250-B Universal Asynchronous Receiver/Transmitter," 1990, pp. 4-3 to 4-18; Connection Diagram, p. 4-3.

7. Reference for Figure 6.2: Ibid. 5.0 Block diagram, p. 4-10, Connection diagrams, p. 4-12.

8. Reference for Figure 6.3: National Semiconductor, Data Sheet, *INS8250-B Asynchronous Communications Element,* "INS8250, INS8250-B Universal Asynchronous Receiver/Transmitter," May 1980, Figure 1, p. 15.

9. Reference for Figure 6.4: National Semiconductor Corporation, *Data Communications Local Area Networks/UARTS,* 1990, pp. 4-3 to 4-18; 9.0 Typical Applications, p. 4-17.

10. Reference for Figure 6.5: Ibid. 8.2 Typical Clock Circuits, p. 4-14.

11. Reference for Figure 6.6: Ibid. 3.0 AC Electrical Characteristics, Baud Generator p. 4-6 (NS16450 p. 4-22); 4.0 Timing Waveforms, /BAUDOUT Timing, p. 4-7.

12. Reference for Figure 6.7: Ibid. 3.0 AC Electrical Characteristics, p. 4-6 (NS16450 p. 4-22); 4.0 Timing Waveforms, Read and Write Timing, p. 4-8.

13. Reference for Figure 6.8: Ibid. 3.0 AC Electrical Characteristics, Receiver p. 4-6, Transmitter p. 4-7, (NS16450 p. 4-22, 4-23); 4.0 Timing Waveforms, Receiver and Transmitter Timing, p. 4-9.

14. Reference for Figure 6.9: Ibid. 3.0 AC Electrical Characteristics, Modem Control p. 4-7, (NS16450 p. 4-23); 4.0 Timing Waveforms, MODEM Controls Timing, p. 4-9.

15. Reference for Figure 6.10: Ibid. 9.0 Typical Applications, p. 4-18.

16. Reference for Table 6.1: Ibid. 6.0 Pin Descriptions, Table I, "Register Addresses," p. 4-11.

17. Reference for Table 6.2: Ibid. Table 6.III, "Baud Rates Using 1.8432 MHz Crystal," and Table 6.IV, "Baud Rates Using 3.072 MHz Crystal," p. 4-14.

18. Reference for Table 6.4: Ibid. Table II, "Summary of Registers," p. 4-13.

19. Reference for Table 6.5: Ibid. 8.1 Line Control Register, p. 4-13.

20. Reference for Table 6.6: Ibid. Table V, "Interrupt Control Functions," p. 4-15.

21. Reference for Table 6.6: Scott Mueller, *Upgrading and Repairing PCs*, Que, 8th Edition, Table 11.4, "UART Chips in PC or AT Systems," p. 434.

Chapter 7
The NS16550/NS16550A/
NS16550AF UART With FIFOs

Introduction

The NS16550/A/AF UART[1] closely resembles the NS16450 UART described in Chapter 6, being basically an improved version of that device[2]. The NS16550/A/AF incorporates the existing registers of the NS16450 and can run all existing IBM PC, XT, AT, RT, and compatible serial port software.

In addition, the device has a programmable mode that incorporates new high-performance features. (For reasons described later, the NS16550 must not be operated in the FIFO mode.) Increases in bus speed and specialized functions provide faster timing with respect to hardware and more efficient use of system software. The NS16550/A/AF UART and the NS16450 UART—as they are described in this book—are functionally interchangeable upon power-up, with two pin usage changes shown in the packaging and block diagram of Figure 7.2 and with the wiring of the test setup.

Figure 7.1 illustrates the basic MICROBUS[3] configuration.

Two NS16550/A/AF packages are available: the 40-pin DIP and a 44-pin PCC, as shown in *Figure 7.2.* The block diagram included provides an overview of the device's internal organization.

Figure 7.3 illustrates an 8088-computer application. This diagram may be helpful to our understanding as we perform the bench-top operations later.

The basic functions of the device are: (1) to accept parallel input from the data bus, perform a serial-to-parallel conversion, and output it in serial format, and (2) to receive serial input and to present it to the CPU bus in a parallel format. These features are enhanced by the addition of transmitter and receiver first-in first-out (FIFO) memories. A wide range of baud rates is available internally, as are the communication protocols required for modem operation.

As we see in the application drawings, the clock source may be linked to that for the system. There is also an alternative crystal source option seen in *Figure 7.4.* The baud rates obtainable using one of three crystal frequencies are described in *Tables 7.3* and *7.4.* Related BAUDOUT waveforms are illustrated in *Figure 7.5.*

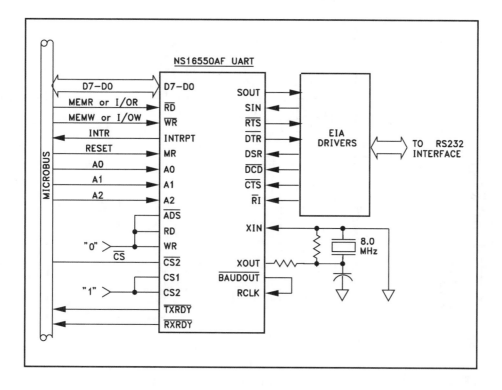

Figure 7.1. The NS16550A/AF MICROBUS connection diagram.

Timing diagrams for read and write cycles are shown in *Figure 7.6*. This figure includes the AC characteristics. The data is shown for reference only, as speed with our bench-top operation is irrelevant. But the diagrams themselves are helpful in observing the operational sequences required. If timing data is important to your need, the manufacturer's data should be consulted.

Timing diagrams for receiver and transmitter functions are shown in *Figure 7.7*. This figure also includes the AC characteristics.

Timing diagrams for modem controls are shown in *Figure 7.8*, which also includes the AC characteristics.

Timing diagrams for the FIFO receiver and transmitter timing are shown in *Figures 7.9* and *7.10*, respectively.

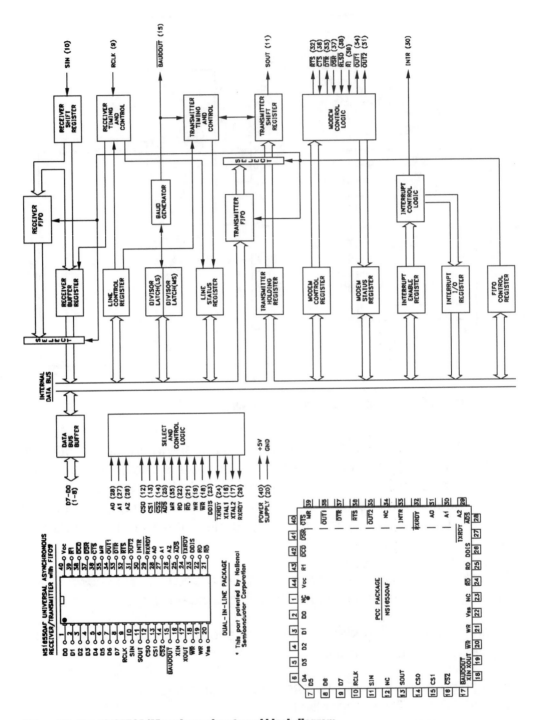

Figure 7.2. The NS16550A/AF package pinouts and block diagram.

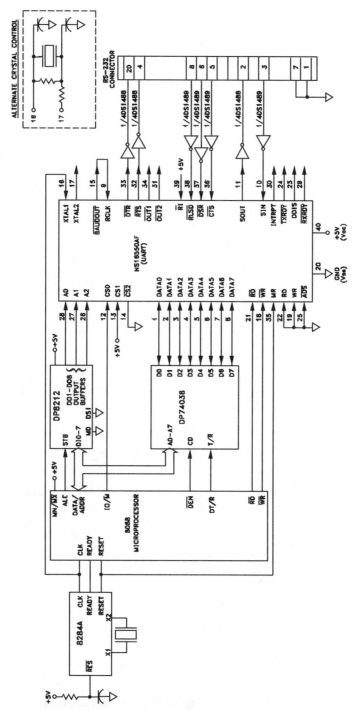

Figure 7.3. Application of an IN16550A/AF to an 8088 CPU.

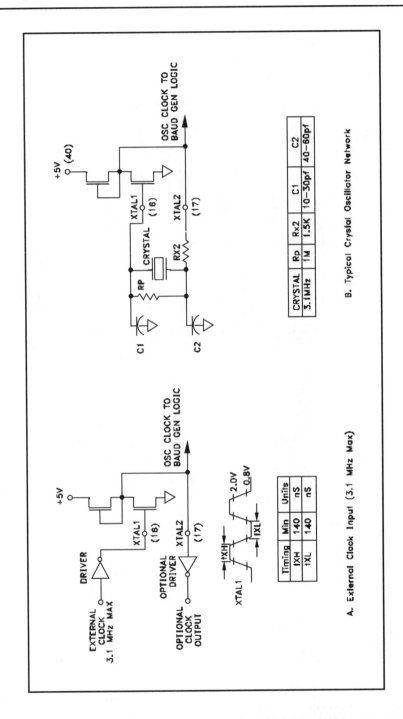

Figure 7.4. Clock applications: (A) connection diagram for an external clock source; (B) a typical crystal oscillator network.

BAUD RATES USING 1.8432 MHz CRYSTAL

Desired Baud Rate	Divisor Used to Generate 16x Clock	Percent Error Difference Between Desired & Actual
50	2304	—
75	1536	—
110	1047	0.026
134.5	857	0.058
150	768	—
300	384	—
600	192	—
1200	96	—
1800	64	—
2000	58	0.69
2400	48	—
3600	32	—
4800	24	—
7200	16	—
9600	12	—
19200	6	—
38400	3	—
56000	2	2.86

NOTE: 1.8432 MHz is the standard 8080 frequency divided by 10.

BAUD RATES USING 3.072 MHz CRYSTAL

Desired Baud Rate	Divisor Used to Generate 16x Clock	Percent Error Difference Between Desired & Actual
50	3840	—
75	2560	—
110	1745	0.026
134.5	1428	0.034
150	1280	—
300	640	—
600	320	—
1200	160	—
1800	107	0.312
2000	96	—
2400	80	—
3600	53	0.628
4800	40	—
7200	27	1.23
9600	20	—
19200	10	—
38400	5	—

Table 7.3. Baud rates obtained by divisors used to generate 16X clocks using a crystal frequency of 1.8432 MHz, and a crystal frequency of 3.072 MHz. Coordinate this table with the text provided under the heading "Programmable Baud Generator ."

BAUD RATES USING 8 MHz CRYSTAL

Desired Baud Rate	Divisor Used to Generate 16x Clock	Percent Error Difference Between Desired & Actual
50	10000	——
75	6667	0.005
110	4545	0.010
134.5	3717	0.013
150	3333	0.010
300	1667	0.020
600	833	0.040
1200	417	0.080
1800	277	0.080
2000	250	——
2400	208	0.160
3600	139	0.080
4800	104	0.160
7200	69	0.644
9600	52	0.160
19200	26	0.160
38400	13	0.160
56000	9	0.790
128000	4	2.344
256000	2	2.344

Table 7.4. Baud rates obtained by divisors used to generate 16X clocks based on a crystal frequency of 8 MHz.

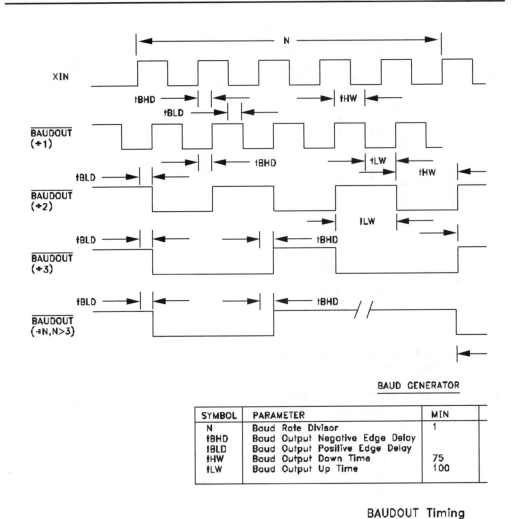

BAUD GENERATOR

SYMBOL	PARAMETER	MIN
N	Baud Rate Divisor	1
tBHD	Baud Output Negative Edge Delay	
tBLD	Baud Output Positive Edge Delay	
tHW	Baud Output Down Time	75
tLW	Baud Output Up Time	100

BAUDOUT Timing

Figure 7.5. BAUDOUT timing diagram with generator representative times. *(Continued on next page.)*

Figure 7.11 illustrates an application of the driver disable output, pin 23 (DDIS), in disabling or controlling the direction of a data bus transceiver between the CPU and the data bus.

The remaining figures relate to the bench-top operation and are described in that section of the chapter.

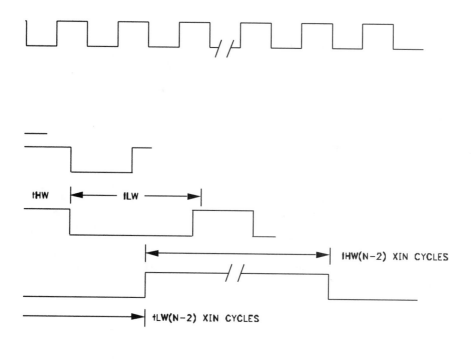

MAX	UNITS	CONDITIONS
$2^{16}-1$		
175	nS	100pf Load
175	nS	100pf Load
	nS	*
	nS	*

* fx=8.0MHz,– 2, 100pf load

Table 7.1 relates the address inputs to the register they select. We find 11 registers residing within the block diagram of Figure 7.2. Table 7.1 defines the addressing for these and the divisor-latch access bit (DLAB).

Table 7.2 describes the master reset functions, about which we will learn more later.

Table 7.3 definse available baud rates with a 1.8432-MHz and 3.072-MHz crystal. Table 7.4 extends the crystal frequency to 8.0 MHz.

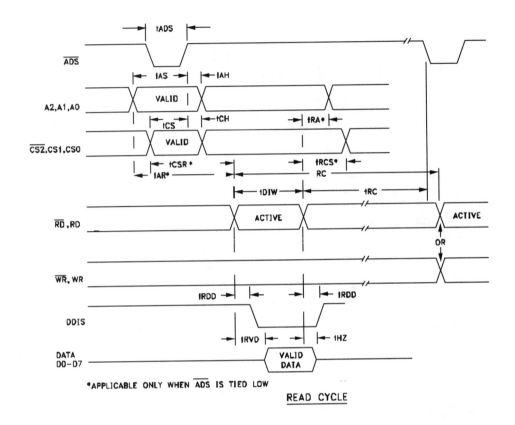

*APPLICABLE ONLY WHEN ADS IS TIED LOW

READ CYCLE

NS16550AF AC

SYMBOL	PARAMETER	MIN	MAX	UNITS	CONDITIONS
tADS	Address Strobe Width	60		nS	
tAS	Address Setup Time	60		nS	
tAH	Address Hold Time	0		nS	
tAR	RD,RD Delay from Address	30		nS	Note 1
tAW	WR,WR Delay from Address	30		nS	Note 1
tCH	Chip Select Hold Time	0		nS	
tCS	Chip Select Setup Time	60		nS	
tCSR	RD,RD Delay from Chip Select	30		nS	Note 1
tCSW	WR,WR Delay from Chip Select	30		nS	Note 1
tCH	Data Hold Time	30		nS	
tDS	Data Setup Time	30		nS	
tHZ	RD,RD to Floating Data Delay	0	100	nS	@100 pf, Note 3
tMR	Master Reset Pulse Width	5		nS	
tRA	Address Hold Time from RD,RD	20		nS	Note 1
tRC	Read Cycle Delay	125		nS	

Figure 7.6. Timing diagram for the NS16550A/AF Read and Write Cycles with AC characteristics. *(Continued on next page.)*

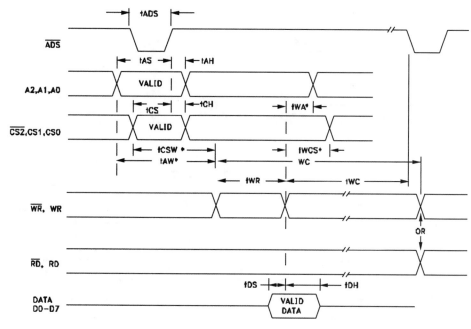

*APPLICABLE ONLY WHEN $\overline{ADS}$ IS TIED LOW

WRITE CYCLE

CHARACTERISTICS

SYMBOL	PARAMETER	MIN	MAX	UNITS	CONDITIONS
tRCS	Chip Select Hold Time from RD,DR	20		nS	Note 1
tRD	$\overline{RD}$,RD Strobe Width	125		nS	
tRDD	$\overline{RD}$,RD to Driver Enable/Disable		60	nS	@100pf, Note 3
tRVD	Delay from $\overline{RD}$,RD to Data		125	nS	@00 pf Loading
tWA	Address Hold Time from $\overline{WR}$,WR	20		nS	Note 1
tWC	Write Cycle Delay	150		nS	
tWCS	Chip Select Hold Time from $\overline{WR}$,WR	20		nS	Note 1
tWR	$\overline{WR}$,WR Strobe Width	100		nS	
tXH	Duration of Clock High Pulse	55		nS	Ext.Clk, 8MHz Max
tXL	Duration of Clock Low Pulse	55		nS	Ext.Clk, 8MHz Max
RC	Read Cycle = tAR+tRD+tRC	280		nS	
WC	Write Cycle = tAW+tWR+tWC	280		nS	

NOTE 1. Applicable only when $\overline{ADS}$ is tied Low.
NOTE 3. Charge and discharge time is determined by VOL, VOH and the external loading

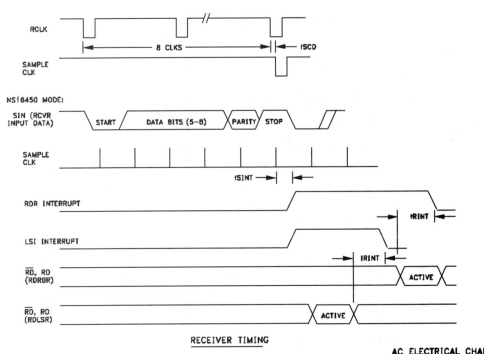

RECEIVER TIMING

AC ELECTRICAL CHAR

SYMBOL	PARAMETER
RECEIVER	
tSCD	Delay from RCLK to Sample Time
tSINT	Delay from Stop to Set Interrupt
tRINT	Delay from RD,RD (RD RBR/or RD LSR) to Reset Interrupt
tRXI	Delay from RD RBR to RXRDY inactive
TRANSMITTER	
tHR	Delay from WR,WR (WR THR) to reset Interrupt
tIRS	Delay from Initial INTR Reset to Transmit
tSI	Delay from Initial Write to Interrupt
tSTI	Delay from Stop to Interrupt (THRE)
tIR	Delay from RD,RD (RD IIR) to Reset Interrupt (THRE)
tSXA	Delay from Start to TXRDY active
tWXI	Delay from Write to TXRDY active

Figure 7.7. Timing diagram for the NS16550A/AF Receiver and Transmitter Timing with AC characteristics. *(Continued on next page.)*

Table 7.5 provides a summary of the register addressing and their functions. These registers control all of the device operations, including the transmission and reception of data. The registers are accessible from the CPU for their operation, which, in this book, is us!

Table 7.6 shows the four-bit code selection for the word length range of five to eight bits. The remaining functions of the line control register are shown in Table 7.5. Similarly, *Table 7.8* shows the bit coding for the

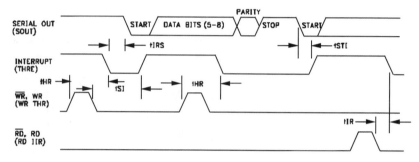

TRANSMITTER TIMING

IARACTERISTICS

	MIN	MAX	UNITS	CONDITIONS
to		2 typ	uS	
		1 typ	uS	Note 2
		1 typ	uS	100 pf Load
		290	nS	
		175	nS	100 pf Load
'I Start	8	24	BAUDOUT Cycles	
	16	24	BAUDOUT Cycles	Note 1
	8	8	BAUDOUT Cycles	Note 1 / 100 pf Load
		8	BAUDOUT Cycles	100 pf Load
		195	nS	100 pf Load

NOTE 1. This delay will be lengthened by one character time, minus the last stop bit time if transmitter interrupt delay circuit is active. (See FIFO Interrupt Mode Operation).

receiver FIFO bit selection. Table 7.5 shows the overall functions of the FIFO control register.

Table 7.9 provides the addressing and function descriptions for the interrupt control functions.

The remaining tables relate to the bench-top operation and are described in that section of the chapter.

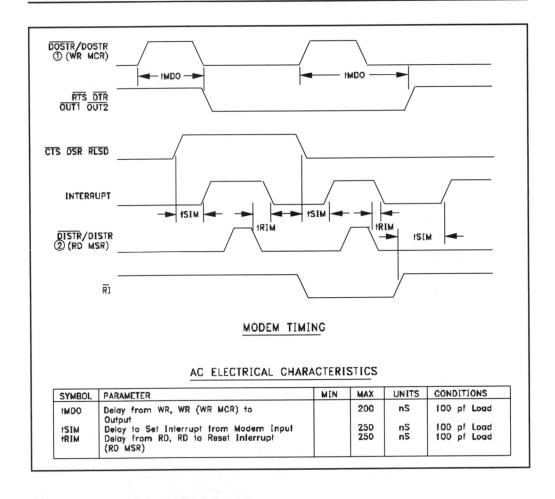

MODEM TIMING

AC ELECTRICAL CHARACTERISTICS

SYMBOL	PARAMETER	MIN	MAX	UNITS	CONDITIONS
tMDO	Delay from WR, WR (WR MCR) to Output		200	nS	100 pf Load
tSIM	Delay to Set Interrupt from Modem Input		250	nS	100 pf Load
tRIM	Delay from RD, RD to Reset Interrupt (RD MSR)		250	nS	100 pf Load

Figure 7.8. The diagram for the NS16550A/AF MODEM Timing with AC characteristics.

REGISTER SELECT TABLE

DLAB	A2	A1	A0	REGISTER
0	0	0	0	Receiver Buffer (Read) Transmitter Holding Register (Write)
0	0	0	1	Interrupt Enable
X	0	1	0	Interrupt ID (Read Only)
X	0	1	1	FIFO Control (Write)
X	0	1	1	Line Control
X	1	0	0	MODEM Control
X	1	0	1	Line Status
X	1	1	0	MODEM Status
X	1	1	1	Scratch
1	0	0	0	Divisor Latch (LSB)
1	0	0	1	Divsior Latch (MSB)

Table 7.1. The DLAB Register Select addressing. Note that all of these registers are contained within the Divisor Latch Access and communicate with the CPU via the 8-bit Data Bus. The DLAB is described in Table 7.5.

REGISTER/SIGNAL	RESET CONTROL	RESET STATE
Interrupt Enable Register	Master Reset	All Bits Low
		(0-3 Forced, 4-7 Permanent)
Interrupt Identification Register	Master Reset	Bit 0 is High, Bits 1 and 2 Low
		Bits 3-7 are Permanently Low
Line Control Register	Master Reset	All Bits Low
MODEM Control Register	Master Reset	All Bits Low
Line Status Register	Master Reset	All Bits Low
		Except Bits 5,6 are High
MODEM Status Register	Master Reset	Bits 0-3 Low
		Bits 4-7 – Input Signal
SOUT	Master Reset	High
INTRPT (RCVR Errs)	Read LSR/MR	Low
INTRPT (RCVR Data Ready)	Read RBR/MR	Low
INTRPT (THRE)	Read IIR/Write THR/MR	Low
INTRPT (Modem Status Changes)	Read MSR/MR	Low
OUT2	Master Reset	High
RTS	Master Reset	High
DTR	Master Reset	High
OUT1	Master Reset	High
RCVR FIFO	MR/FCR1●FCR0/△FCR0	All bits low
XMIT FIFO	MR/FCR1●FCR0/△FCR0	All bits low

Table 7.2. UART Reset Functions. Observe that each DLAB Register is unique with respect to the Master Reset.

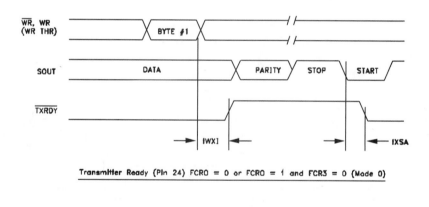

Transmitter Ready (Pin 24) FCR0 = 0 or FCR0 = 1 and FCR3 = 0 (Mode 0)

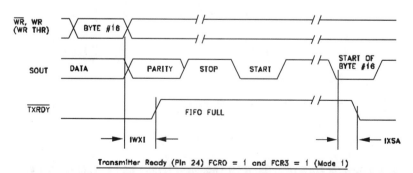

Transmitter Ready (Pin 24) FCR0 = 1 and FCR3 = 1 (Mode 1)

Figure 7.10. Diagram for the NS16550A/AF FIFO Transmitter Timing with AC characteristics.

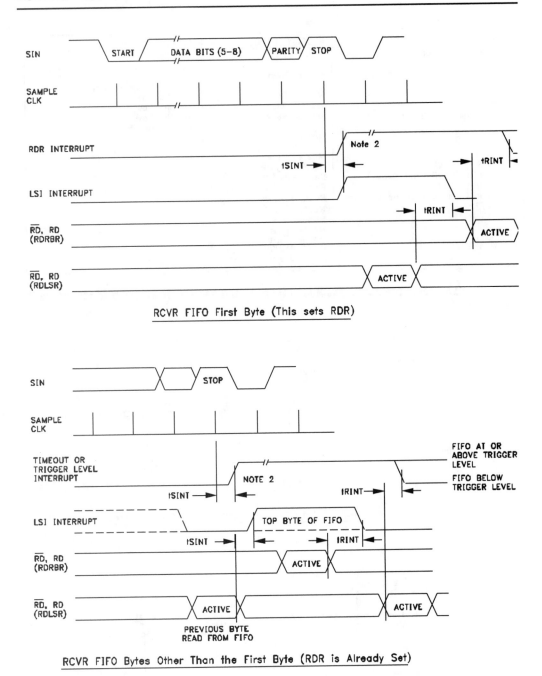

Figure 7.9. The diagram for the NS16550A/AF FIFO Receiver Timing with AC characteristics. *(Continued on next page.)*

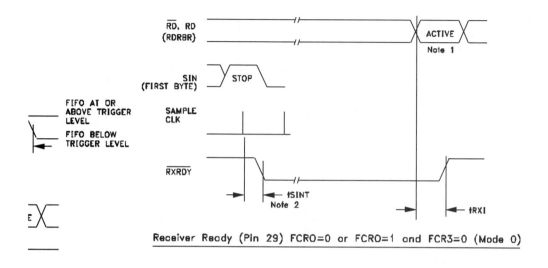

Receiver Ready (Pin 29) FCRO=0 or FCRO=1 and FCR3=0 (Mode 0)

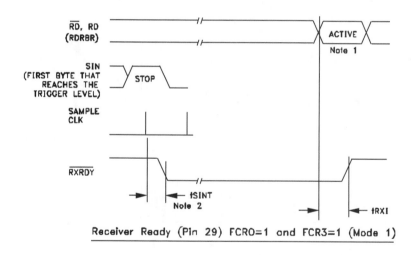

Receiver Ready (Pin 29) FCRO=1 and FCR3=1 (Mode 1)

Note 1: This is the reading of the last byte in the FIFO.
Note 2: If FCRO=1, then tSINT=3 RCLKs. For a timed interrupt, tSINT=8 RCLKs.

SUMMARY OF REGISTERS

Bit No.	REGISTER ADDRESS											
	0 DLAB=0	0 DLAB=0	1 DLAB=0	2	2	3	4	5	6	7	0 DLAB=1	1 DLAB=1
	Receiver Buffer Register (Read Only) RBR	Transmitter Holding Register (Write Only) THR	Interrupt Enable Register IER	Interrupt Identification Register (Read Only) IIR	FIFO Control Register (Write Only) FCR	Line Control Register LCR	MODEM Control Register MCR	Line Status Register LSR	MODEM Status Register MSR	Scratch Register SCR	Divisor Latch (LS) DLL	Divisor Latch (MS) DLM
0	Data Bit 0 Note 1	Data Bit 0	Enable Received Data Available Interrupt (ERBFI)	"0" if Interrupt Pending	FIFO Enable	Word Length Select Bit 0 (WLS0)	Data Terminal Ready (DTR)	Data Ready (DR)	Delta Clear to Send (DCTS)	Bit 0	Bit 0	Bit 8
1	Data Bit 1	Data Bit 1	Enable Transmitter Holding Register Empty Interrupt (ETBEI)	Interrupt ID Bit (0)	RCVR FIFO Reset	Word Length Select Bit 1 (WLS1)	Request to Send (RTS)	Overrun Error (OR)	Delta Data Set Ready (DDSR)	Bit 1	Bit 1	Bit 9
2	Data Bit 2	Data Bit 2	Enable Receiver Line Status Interrupt (ELSI)	Interrupt ID Bit (1)	XMIT FIFO Reset	Number of Stop Bits (STB)	OUT 1	Parity Error (PE)	Trailing Edge Ring Indicator (TERI)	Bit 2	Bit 2	Bit 10
3	Data Bit 3	Data Bit 3	Enable MODEM Status Interrupt (EDSSI)	Interrupt ID Bit (2) Note 2	DMA Mode Select	Parity Enable (PEN)	OUT 2	Framing Error (FE)	Delta Receiver Line Signal Detect	Bit 3	Bit 3	Bit 11
4	Data Bit 4	Data Bit 4	0	0	Reserved	Even Parity Select (EPS)	Loop	Break Interrupt (BI)	Clear To Send (CTS)	Bit 4	Bit 4	Bit 12
5	Data Bit 5	Data Bit 5	0	0	Reserved	Stick Parity	0	Transmitter Holding Register Empty (THRE)	Data Set Ready (DSR)	Bit 5	Bit 5	Bit 13
6	Data Bit 6	Data Bit 6	0	FIFOs Enabled Note 2	RCVR Trigger LSB	Set Break	0	Transmitter Shift Register Empty (TSRE)	Ring Indicator (RI)	Bit 6	Bit 6	Bit 14
7	Data Bit 7	Data Bit 7	0	FIFOs Disabled Note 2	RCVR Trigger MSB	Divisor Latch Access Bit (DLAB)	0	Error in RCVR FIFO Note 2	Data Carrier Detect (DCD)	Bit 7	Bit 7	Bit 15

Note 1. Bit 0 is the least significant bit. It is the first bit serially received or transmitted.
Note 2. These bits are always 0 in the NS16450 mode.

Table 7.5. A summary of the Register functions accessible through the Divisor Latch Access Bit (DLAB) and the Register Addresses. It will be helpful to relate this table to the block diagram of Figure 7.2.

LINE CONTROL REGISTER

BIT 1	BIT 0	WORD LENGTH
0	0	5 Bits
0	1	6 Bits
1	0	7 Bits
1	1	8 Bits

Table 7.6. This table defines the DLAB bit coding for the data word length as coded for serial transmission. Note that there is a relationship between word length and the number of Stop bits.

Table 7.8. This table defines the DLAB bit coding for the Interrupt Set and Reset Functions within the Interrupt Identification Register (IIR). Note this is a Read Only register.

FIFO Mode Only	Interrupt Identification Register			Interrupt Set and Reset Functions			
Bit 3	Bit 2	Bit 1	Bit 0	Prority Level	Interrupt Type	Interrupt Source	Interrupt Reset Control
0	0	0	1	–	None	None	
0	1	1	0	Highest	Receiver Line Status	Overrun Error or Parity Error or Framing Error or Break Interrupt	Reading the Line Status Register
0	1	0	0	Second	Received Data Available	Receiver Data Available or Trigger Level reached	Reading the Receiver Buffer Register or the FIFO drops below the Trigger Level
1	1	0	0	Second	Character Timeout Indication	No characters have been removed or input to the RCVR FIFO during the last 4 char times and there is at least 1 char in it during this time.	Reading the Receiver Buffer Register
0	0	1	0	Third	Transmitter Holding Register Empty	Transmitter Holding Register Empty	Reading the IIR Register (If Interrupt Source) or Writing into the Transmitter Holding Register
0	0	0	0	Fourth	MODEM Status	Clear to Send or Data Set Ready or Ring Indicator or Data Carrier Detect	Reading the Modem Status Register

HOW TO USE THE DATA ENTRY R/W SWITCH WITH THE NS16550AF

With this IC data entry on the bus is not related to the clock.

1. To write to the Data Bus:
   Ensure the Data Entry R/$\overline{W}$ switch is in the R position.
   Establish the Data Pattern on the Data Entry Module
2. Use the Address Select DIP switch to enter the required address.
3. Toggle the Data Entry R/$\overline{W}$ switch to the $\overline{W}$ position.
   Observe the pattern appears on LED Module 4
4. Toggle the 16550 Write switch.
5. Return the Data Entry R/$\overline{W}$ switch to the R position.
   Unless noted otherwise a successful write will be followed by
   a display of the entered data with the switch in the R position.

Table 7.9. Instructions on usage of the Data Entry and R/W toggle switches found on the Bus Entry Module.

Figure 7.11. Application of DDIS to disable communications between the Data Bus and the UART.

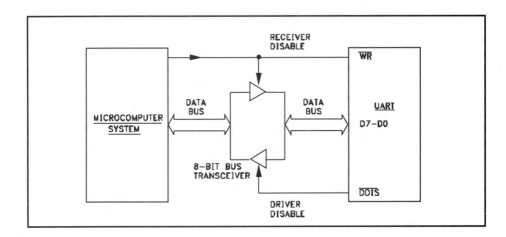

UART Pin Descriptions

The function descriptions that follow are in alphabetical order by the identifying acronym. The corresponding pin numbers immediately follow, preceding a summary of the function. Note that in this book, "/" defines the complement—that is, the term is active low.

Function	Pin(s)	Description
A2,A1,A0	26-28	Signals applied to these three inputs select a UART register for reading from or writing to by the CPU during a transfer of data. Table 6.1 relates the address inputs to the register they select. The state of the divisor latch access bit (DLAB), the most significant bit of the line control register, determines the slection of certain UART registers. The DLAB (Bit 7 of address 011) must be set high to access the baud-generator divisor latches. Note that an active /ADS input is required when the select addresses are not stable during the duration of a read or write operation. When not a factor, this input line is tied permanently low.
/BAUDOUT	15	Baud Out Pin. This is the 16X clock signal provided to the transmitter section of the UART. The clock rate is equal to the main reference oscillator frequency divided by the specified divisor in the baud-generator divisor latches. This clock signal may also be utilized by the receiver section by connection to the RCLK input, Pin 9.
CS0,CS1,	12-14	The chip is selected with CS0,CS1 high and /CS2 low.
/CS2		This enables communication between the UART and the CPU. The positive edge of an active address strobe signal latches the decoded chip select signals to complete chip selection. If /ADS is always low, the chip selects should stabilize according to the tCSW schedule (see Figure 7.6).
CSOUT	24	Chip Select Out. When this line is high, it indicates the chip has been selected by active CS0,CS1,/CS2 inputs. This line must be at logic one to enable a data transfer. The line goes low with the chip deselected.
/CTS	36	Clear-to-Send. When this pin is low, it indicates the modem or data set is ready to exchange data. This signal is a modem status input whose condition can be tested by the CPU reading Bit 4 (CTS) of the modem status register. Bit 4 is the complement of the /CTS signal. Bit 0 (DCTS) of the modem status register indicates whether the /CTS input has changed status since the previous

reading of the modem status register. /CTS has no effect on the transmitter. Note that whenever the CTS bit of the modem status register changes state, an interrupt is generated if the modem status interrupt is enabled.

D7-D0   1-8   The data bus is comprised of eight TRI-STATE[3] input/output lines. The bus provides bidirectional communications between the UART and the CPU. Data, control words, and status information are transferred via this bus.

/DCD   38   Data Carrier Detect pin. When this pin is low, it indicates the data carrier has been detected by the modem or data set. The /DCD signal is a modem status input whose condition can be tested by the CPU reading Bit 7 (DCD) of the modem status register. Bit 7 is the complement of the /DCD signal. Bit 3 (DDCD) of the modem status register indicates whether the /DCD input has changed state since the previous reading of the modem status register. /DCD has no effect on the receiver. Note that whenever the DCD bit of the modem status register changes state, an interrupt is generated if the modem status interrupt is enabled.

DDIS   23   Driver Disable. This line goes low whenever the CPU is *reading* data from the UART. It can be used to disable or control the direction of a data bus transceiver between the UART and the CPU (refer to Figure 7.11).

/DSR   37   Data Set Ready. When low, this indicates the modem or data set is ready to establish the communication link with the UART. This signal is a modem status input whose condition can be tested by the CPU reading Bit 5 (DSR) of the modem status register. Bit 5 is the complement of the /DSR signal. Bit 1 (DDSR) of the modem status register indicates whether the /DSR input has changed state since previous reading of the modem status register. Note that whenever the DSR bit of the modem status register changes state, an Interrupt is generated if the modem status interrupt is enabled.

/DTR   33   Data Terminal Ready. When low, this indicates the modem or data set is ready to establish the communication link with the UART. The /DTR signal can be set to an active low by programming Bit 0 (DTR) of the modem control register to a high level. A master reset operation sets this signal to its inactive (high) state. Loop-mode operation holds the signal in its inactive state.

INTR   30   Interrupt. This line goes high whenever any one of the following interrupt types has an active high condition and is enabled via the IER: receiver line status, received data available, transmitter-holding register empty, and modem status. The INTR line is reset low upon the appropriate interrupt service or a master reset operation.

| MR | 35 | Master Reset. With this input high all registers are cleared other than the receiver buffer, transmitter-holding register, and the divisor latches. All UART control logic is cleared. The states of outputs SOUT, INTR, /OUT1, /OUT2, /RTS, and /DTR are as shown in Table 7.2. The reset status of the RCVR and XMIT FIFOs are included also. This input is buffered with a TTL compatible Schmitt trigger with typical 0.5V hysteresis. |

| /OUT1 | 34 | Output 1. This is a user-designated output that is set to an active low by programming Bit 2 (OUT 1) of the modem control register to a logic 1 level. A master reset sets this signal to its inactive (high) state. Loop-mode operation holds it in the inactive state. In the XMOS parts, this will achieve TTL logic levels. |

| /OUT2 | 31 | Output 2. This is a user-designated output that is set to an active low by programming Bit 3 (OUT 2) of the modem control register to a logic 1 level. A master reset sets this signal to its inactive (high) state. Loop-mode operation holds it in the inactive state. In the XMOS parts, this will achieve TTL logic levels. |

| RCLK | 9 | Receiver Clock. This input is the 16 X baud-rate clock for the receiver portion of the chip. |

| RD,/RD | 22,21 | Read. With RD high, or /RD low, with the chip selected, the CPU can read status information or data from the selected UART register. Note that only an active RD or /RD is required to transfer data during a read operation. Therefore, tie either the RD line permanently low or the /RD permanently high when not being used. |

| /RI | 39 | Ring Indicator. When low, this input indicates that a telephone ringing signal has been received by the modem or data set. The /RI signal is a modem status input whose condition can be tested by the CPU reading Bit 6 (RI) of the modem status register. Bit 6 is the complement of the /RI signal. Bit 2 (TERI) of the modem status register indicates whether the /RI input signal has changed from a low to a high state since the previous reading of the modem status register. Note that whenever the RI bit of the modem status register changes from a high to a low state, an Interrupt is generated if the modem status interrupt is enabled. |

| /RTS | 32 | Request-to-Send. When low, this output line informs the modem or data set that the UART is ready to exchange data. The /RTS signal can be set to an active low by programming Bit 1 (RTS) of the modem control register to a high level. A master reset operation sets this signal to its inactive (high) state. Loop-mode operation holds the signal in its inactive state. |

SIN	10	Serial Input. Serial data input from the communication link: modem, data set, or peripheral device.
SOUT	11	Serial Out. The composite serial data output to the communication link: modem, data set, or peripheral device. The SOUT signal is set to the marking (logic 1) state upon a master reset or when the transmitter is idle.
/TXRDY,	24,29	These pin connections provide for transmitter and receiver /RXRDY DMA signaling. When operating in the FIFO mode, one of two types of DMA signaling per pin can be selected via FCR3. When operating in the NS16450 mode, only DMA Mode 0 is allowed, which supports single transfer DMA in which transfer is made between bus cycles. Mode 1 supports multitransfer DMA, in which multiple transfers are made continuously until the RCVR FIFO has been emptied or the XMIT FIFO has been filled.
RXRDY	29	Mode 0: When in the NS16450 mode (FCR0=0) or in the FIFO mode (FCR0=1, FCR3=0) and there is at least one character in the RCVR FIFO or RCVR holding register, the RXRDY pin will be low active. Once it is activated, the RXRDY pin will go inactive when there are no more characters in the FIFO or holding registers.
RXRDY	29	Mode 1: When in the FIFO mode (FCR0=1) with FCR3=1 and the trigger level or the timeout has been reached, the RXRDY pin will go low inactive. Once it is activated, it will go inactive when there are no more characters in the FIFO or holding register.
TXRDY	24	Mode 0: When in the NS16450 mode (FCR0=0) or the FIFO mode (FCR0=1, FCR3=0) with no more characters in the XMIT FIFO or XMIT holding register, the TXRDY pin will be low inactive. Once it is activated, the TXRDY pin will go inactive after the first character is loaded in the XMIT FIFO or holding register.
TXRDY	24	Mode 1: When in the FIFO mode (FCR0=1) with FCR3=1 and no more characters in the XMIT FIFO, the /TXRDY pin go low active. This pin will become inactive when the XMIT FIFO is completely full.
Vcc	40	+5V dc supply.
Vss	20	Ground (0V) reference.
WR,/WR	19,18	Write. With WR high, or /WR low, with the chip selected, the CPU can write control words or data into the selected UART register. Note that only an active WR or /WR is required to transfer data to the UART during a write operation. Therefore, tie either the WR line permanently Low or the /WR permanently high, when not being used.

XIN	16	External Crystal Input. This signal input is used in conjunction with XOUT to form a feedback circuit for the baud-rate generator's oscillator. If a clock signal will be generated off-chip, it should drive the baud-rate generator through this pin.
XOUT	17	External Crystal Output. This signal input is used in conjunction with XIN to form a feedback circuit for the baud-rate generator's oscillator. If a clock signal will be generated off-chip, this pin is unused.

UART Registers

Table 7.5 presents a summary of the UART registers. Register access and their content are identical for all versions of the device. These are all accessible to the CPU via the eight-bit data bus. Control of the device operation is exercised via these register functions. If you find this table confusing, you're not alone.

Return to the register addresses given in Table 7.1. Let's look at the leftmost column with the heading DLAB. Note that addresses 000 and 001 appear twice, at the beginning with DLAB shown as "0", and at the bottom with DLAB shown as "1." When we get to the bench-top procedures this will be clearer, but what is needed, whether here on the bench or with a computer program, is to go to address 011, the line control register, and set Bit 7 to "1" with all other bits at logic "0." This puts us in the DLAB.

Now we can set the address to 000 and enter the divisor latch least significant bits. Then proceed to address 001 to enter the most significant bits. With that accomplished, we can return to address 011 and set Bits 0 through 6 for our data communications with Bit 7 now set to 0. With this in mind, we can investigate the register functions. Note that resetting the device will not affect the divisor latch settings.

Line Control Register

The line control register, address 011, specifies the format of the asynchronous data communication exchange. Table 7.6 defines the coding of Bits 0 and 1 of this register for selection of the required character length in the exchange.

Bit 2 specifies the number of stop bits transmitted and received in each serial character. If Bit 2 is a logic 0, one stop bit is generated or checked in the transmitted data. If Bit 2 is a logic 1 when a five-bit word length is selected via Bits 0 and 1, one and one-half stop bits are generated. If Bit 2 is a logic 1 whenever one of the other three length options are selected, two stop bits are selected. Note that the receiver checks the first stop bit only-regardless of the number of stop bits selected.

Bit 3 is the parity enable bit. When this bit is a logic 1, a parity bit is transmitted and tested for in the received data between the last data word bit and the stop bit. The parity bit is used to produce either an even or odd number of 1s when the data word bits and the parity bit are summed.

Bit 4 is the even parity select bit. When Bit 3 is a logic 1 and Bit 4 is a logic 0, an odd number of logic 1s is transmitted or checked for. When Bit 3 is a logic 1 and Bit 4 is a logic 1, an even number of logic 1s is transmitted or checked for.

Bit 5 is the stick parity bit. When Bits 3, 4, and 5 are logic 1, the parity bit is transmitted and checked as a logic 0. If Bits 3 and 5 are 1 and Bit 4 is a logic 0, then the parity is transmitted and checked as a logic 1. If Bit 5 is a logic 0, the stick parity is disabled.

Bit 6 is the break control bit. It causes a break condition to be transmitted by the UART. When it's set to a logic 1, the serial output (SOUT) is forced to the spacing (logic 0) state. The break is disabled by clearing Bit 6 to a logic 0. The break control bit acts only on SOUT and has no effect on the transmitter logic.

(Note: This feature enables the CPU to alert a terminal in a computer communications system. If the following sequence is used, no erroneous or extraneous characters will be transmitted because of the break.

1. Load in all 0s, pad character, in response to THRE.
2. Set break after the next THRE.

3. Wait for the transmitter to be idle, (TEMT=1), and clear break when normal transmission has to be restored.

During the break, the transmitter can be used as a character timer to accurately determine the break duration.)

Bit 7 is the divisor-latch access bit (DLAB). It must be set high (logic 1) to access the divisor latches of the baud generator during a read or write operation. It must be set low (logic 0) to access the receiver buffer, the transmitter-holding register, or the interrupt enable register and the line and modem status or control registers, or the FIFO control register.

Programmable Baud Generator

The UART contains a programmable baud-rate generator capable of taking any clock input from DC to 8.0 MHz and dividing it by any divisor from 1 to 2e16-1. Clock circuits are depicted in Figure 7.4. Table 7.3 defines available baud rates with a 1.8432-MHz and 3.072-MHz crystal. Table 7.4 provides similar data for an 8.0-MHz crystal.

The output frequency of the baud-rate generator is 16X the baud [divisor # = (frequency input) ÷ (baud rate X 16)]. Two eight-bit latches store the divisor in a 16-bit binary format. These divisor latches must be loaded during initialization in order to ensure proper operation of the baud generator. Upon loading either of the divisor latches, a 16-bit baud counter is immediately loaded.

Line Status Register

This eight-bit register provides status information to the CPU concerning the data transfer. Table 7.5 shows the content of each register bit as follows.

• Bit 0 is the receiver data ready (DR) indicator. This bit is set to a logic 1 whenever a complete incoming character has been received and transferred into the receiver-buffer register. This bit is reset to a logic 0 by reading the data in the receiver buffer register.

• Bit 1 is the overrun error (OE) indicator. This bit indicates that data in the receiver buffer register was not read by the CPU before the next character was transferred into the receiver-buffer register, thereby destroying the previous character. The OE indicator is set to a logic 1 upon detection of an overrun condition and is reset to a logic 0 whenever the CPU reads the content of the line status register.

• Bit 2 is the parity error (PE) indicator. This bit indicates that the received data character does not have the correct even or odd parity as selected by the even-parity select bit. The PE is set to a logic 1 upon detection of a parity error and is reset to a logic 0 whenever the CPU reads the content of the line status register.

• Bit 3 is the framing error (FE) indicator. This bit indicates that the received data did not have a valid stop bit. This bit is set to a logic 1 whenever the stop bit following the last data bit or parity bit is a logic 0 (spacing level).

The FE indicator is reset to a logic 0 whenever the CPU reads the content of the line status register. The UART will try to resynchronize after a framing error. To do this, it assumes that the framing error was due to the next start bit, so it samples this "start" bit twice and then takes in the "data."

• Bit 4 is the break interrupt (BI) indicator. This bit is set to a logic 1 whenever the received data input is held in the spacing (logic 0) state for longer than a full word transmission time-that is, the total time of start bit + data bits + parity + stop bits. The BI indicator is reset to a logic 0 whenever the CPU reads the content of the line status register. Restarting after a break is received requires the SIN pin to be logical 1 for at least ½ bit time.

(Note: Bits 1 through 4 are the error conditions that produce a receiver line status interrupt whenever any of the corresponding conditions are detected and the interrupt is enabled.)

• Bit 6 is the transmitter empty (TEMT) indicator. This bit is set to a logic 1 whenever the transmitter-holding register (THR) and the transmitter shift register (TSR) are both empty. It is reset to logic 0 whenever either the THR or the TSR contains a data character.

• Bit 7 is permanently set to logic 0.

(Note: The line status register is intended for read operations only. Writing to this register is not recommended, as this operation is only used for factory testing. In the FIFO mode, the software must load a data byte in the RxFIFO via loopback mode in order to write to LSR2-LSR4. LSR0 and LSR7 can't be written to in FIFO mode.)

The FIFO Control Register (Write Only)

The functions of the FIFO control register are shown in Table 7.5. This is a write-only register at the same location of the interrupt identification register (IIR), which is read only. This register is employed to enable the FIFOs, clear the FIFOs, set the RCVR FIFO trigger level, and select the type of DMA signaling.

Bit 0: Writing a 1 to FCR0 enables both the XMIT and RCVR FIFOs. Resetting FCR0 will clear all bytes in both FIFOs. When changing from FIFO mode to NS16450 mode and vice versa, data is automatically cleared from the FIFOs. This bit must be a 1 when other FCR are written to or they will not be programmed.

Bit 1: Writing a 1 to FCR2 clears all bytes in the RCVR FIFO and resets its counter logic to 0. The shift register is not cleared. The 1 that is written to this bit position is self-clearing.

Bit 2: Writing a 1 to FCR2 clears all bytes in the XMIT FIFO and resets its counter logic to 0. The shift register is not cleared. The 1 that is written to this bit position is self-clearing.

Table 7.7. This table defines the DLAB bit coding for the RCVR FIFO trigger level.

FIFO CONTROL REGISTER		
Bit 7	Bit 6	RCVR FIFO Trigger Level (Bytes)
0	0	01
0	1	04
1	0	08
1	1	14

Bit 3: Setting FCR3 to a 1 will cause the RXRDY and TXRDY pins to change from mode 0 to mode 1 if FCR0 = 1 (refer to the description of the RXRDY and TXRDY pins).

Bits 4, 5: These bits are reserved for future use.

Bits 6, 7: FCR6 and FCR7 are used to set the trigger level for the RCVR FIFO interrupt, as shown in *Table 7.7.*

Interrupt Identification Register

In order to minimize software overhead during data transfers, the UART prioritizes interrupts into four levels and records these in the interrupt identification register. The four levels of interrupt conditions in order of priorities are receiver line status, received data ready, transmitter-holding register empty, and modem status.

When the CPU accesses the IIR, the UART freezes all interrupts and indicates the highest priority pending interrupt to the CPU. While the CPU access is occurring, the UART records new interrupts but does not change its current indication until the access is complete. Table 7.5 shows the contents of the IIR. Details on each bit follow.

Bit 0 can be used in an interrupt environment to indicate whether an interrupt condition is pending. When this bit is a logic 0, an interrupt is pending and the IIR contents may be used as a pointer to the appropriate interrupt service routine. When this bit is a logic 1, no interrupt is pending.

Bits 1 and 2 are used to identify the highest priority interrupt pending, as indicated in Table 7.8.

Bit 3: In the NS16450 mode, this bit is 0. In the FIFO mode, it is set along with Bit 2 when a timeout interrupt is pending.

Bits 4, 5: These bits are always at logic 0.

Bits 6, 7: These bits are set whenever FCRO = 0.

Interrupt Enable Register

This register enables the five types of UART interrupts. Each interrupt can individually activate the interrupt (INTR) output signal. It is possible to totally disable the interrupt system by resetting Bits 0 through 3 of the interrupt enable register (IER).

Similarly, setting bits of this register to a logic 1 enables the selected interrupt(s). Disabling an interrupt prevents it from being indicated as active in the IIR and from activating the INTR output signal.

All other system functions operate in their normal manner, including the setting of the line status and modem status registers. Table 7.5 shows the contents of the IER. Details of each bit follow.

Bit 0 enables the received data available interrupt (and timeout interrupts in the FIFO mode) when set to logic 1.

Bit 1 enables the transmitter-holding register empty interrupt when set to logic 1.

Bit 2 enables the receiver line status interrupt when set to logic 1.

Bit 3 enables the modem status interrupt when set to logic 1.

Bits 4 through 7 are always logic 0.

Modem Control Register

This register controls the interface with the modem or data set (or a peripheral device emulating a modem). The contents of the modem control register (MCR) are indicated in Table 7.5. Details on each bit follow.

Bit 0 controls the data terminal ready (/DTR) output. When this bit is set to a logic 1, the /DTR output is forced to a logic 0. When this bit is reset, the /DTR output is forced to a logic 1.

(Note: The /DTR output of the UART may be applied to an EIA, inverting the driver-such as the DS1488-to obtain the proper polarity input at the succeeding modem or data set.)

Bit 1 controls the request-to-send (/RTS) output. When this bit is set to a logic 1, the /RTS output is forced to a logic 0. When this bit is reset, the /RTS output is forced to a logic 1.

Bit 2 controls the Output 1 (/OUT1) signal, which is an auxiliary user-designated output. When this bit is set to a logic 1, the /OUT1 output is forced to a logic 0. When this bit is reset, the /OUT1 output is forced to a logic 1.

Bit 3 controls the Output 2 (/OUT2) signal, which is an auxiliary user-designated output. When this bit is set to a logic 1, the /OUT2 output is forced to a logic 0. When this bit is reset, the /OUT2 output is forced to a logic 1.

Bit 4 provides a local feedback feature for diagnostic testing of the UART. When this bit is set to logic 1, the following occur. The transmitter serial output (SOUT) is set to the marking (logic 1) state; the receiver serial input (SIN) is disconnected; the output of the transmitter shift register is "looped back" into the receiver shift register input; the four modem control inputs (/DSR, /CTS, /RI, /DCD) are disconnected; and the four modem control outputs (/DTR, /RTS, /OUT1, /OUT2) are internally connected to the four modem control inputs. The modem control output pins are forced to their inactive state (high). In the diagnostic mode, data that is transmitted is immediately received. This feature allows the processor to verify the transmit-and-received-data paths of the UART. In the diagnostic mode, the transmitter and receiver interrupts are fully operational. The modem control interrupts are also operational, but the interrupts' sources are now the lower four bits of the modem control register instead of the four modem control inputs. The interrupts are still controlled by the interrupt enable register.

Bits 5 through 7 are permanently set to logic 0.

Modem Status Register

This register provides the current state of the control lines from the modem (or peripheral device) to the CPU. In addition to this current-state information, four bits of the modem status register provide change information. These bits are set to a logic 1 whenever a control input from the modem changes state. They are reset to logic 0 whenever the CPU reads the modem status register. Table 7.5 shows the status of the MSR. Details on each bit follow.

Bit 0 is the delta clear-to-send (DCTS) indicator. This bit indicates that the /CTS input to the chip has changed state since the last time it was read by the CPU.

Bit 1 is the delta data set ready (DDSR) indicator. This bit indicates that the /DSR input to the chip has changed state since the last time it was read by the CPU.

Bit 2 is the trailing edge of ring indicator (TERI) detector. This bit indicates that the /RI input to the chip has changed from a low to a high state.

Bit 3 is the delta data carrier detect (DDCD) indicator. This bit indicates that the DCD input to the chip has changed state. (Note: Whenever Bit 0, 1, 2, or 3 is set to logic 1, a modem status interrupt is generated.)

Bit 4 is the complement of the clear-to-send (/CTS) input. If Bit 4 is set to a logic 1, this bit is equivalent to RTS in the MCR.

Bit 5 is the complement of the data set ready (/DSR) input. If Bit 4 is set to a logic 1, this bit is equivalent to DTR in the MCR.

Bit 6 is the complement of the ring indicator (/RI) input. If Bit 4 is set to a logic 1, this bit is equivalent to OUT1 in the MCR.

Bit 7 is the complement of the data carrier detect (DCD) input. If Bit 4 is set to a logic 1, this bit is equivalent to OUT2 in the MCR.

Scratchpad Register

This eight-bit read/write register does not control the UART in any respect. It is intended as a scratchpad register, to be used by the programmer to hold data temporarily.

FIFO Interrupt Mode Operation

When the RCVR FIFO and receiver interrupts are enabled (FCR01=1, IER0=1), the RCVR interrupt will occur as follows.

A. The receive data available interrupt will be issued to the CPU when the FIFO has reached its programmed trigger level; it will be cleared as soon as the FIFO drops its programmed trigger level.

B. The IIR receives data available indication also occurs when the FIFO trigger level is reached and, like the interrupt, it is cleared when the FIFO drops below the trigger level.

C. The receiver line status interrupt (IIR=06), as before, has higher priority than the received data available (IIR=04) interrupt.

D. The data ready bit (LSR0) is set as soon as a character is transferred from the shift register to the RCVR FIFO. It is reset when the FIFO is empty.

When RCVR FIFO and receiver interrupts are enabled, RCVR FIFO timeout interrupts will occur as follows.

A. A timeout interrupt will occur, if the following conditions exist.

• At least one character is in the FIFO.

• The most recent serial character received was longer than four continuous character times ago (if two stop bits are programmed, the second one is included in this time delay).

• The most recent CPU read of the FIFO was longer than four continuous character times ago.

(The maximum time between a received character and a timeout interrupt will be 160 ms at 300 baud with a 12-bit receive character—that is, one start, eight data, one parity, and two stop bits).

B. Character times are calculated by using the RCLK input for a clock signal (this makes the delay proportional to the baud rate).

C. When a timeout interrupt has occurred it is cleared and the timer reset when the CPU reads one character from the RCVR FIFO.

D. When a timeout interrupt has not occurred, the timeout timer is reset after a new character is received or after the CPU reads the RCVR FIFO.

When the XMIT FIFO and transmitter interrupts (FCR0=1, IER1=1), XMIT Interrupts will occur as follows.

A. The transmitter-holding register interrupt (02) occurs when the XMIT FIFO is empty; it is cleared as soon as the transmitter-holding register is written to (1 to 16 characters may be written to the XMIT FIFO while servicing this interrupt) or the IIR is read.

B. The transmitter FIFO empty indications will be delayed 1 character time minus the last stop bit time whenever the following occurs: THRE=1 and there have not been at least two bytes at the same in time in the transmit FIFO, since the last THRE=1. The first transmitter interrupt after changing FCR0 will be immediate if it is enabled.

Character timeout and RCVR FIFO trigger level interrupts have the same priority as the current transmitter-holding register empty interrupt.

FIFO Polled Mode Operation

With FCR0=1 resetting IER0, IER1, IER2, IER3 or all to zero puts the UART in the FIFO polled mode of operation. Since the RCVR and XMITTER are controlled separately either one or both can be in the polled mode of operation. In this mode the user's program will check RCVR and XMITTER status via the LSR. As stated previously:

• LSR0 will be set as long as there is one byte in the RCVR FIFO.
• LSR1 to LSR4 will specify which error(s) has occurred. Character error status is handled the same way as when in the interrupt mode, the IIR is not affected since IER2=0.
• LSR5 will indicate when the XMIT FIFO is empty.
• LSR6 will indicate that both the XMIT FIFO and the shift register are empty.

• LSR7 will indicate whether there are any errors in the RCVR FIFO.

There is no trigger level reached or timeout condition indicated in the FIFO polled mode; however, the RCVR and XMIT FIFOs are still fully capable of holding characters.

NS16550/A/AF and NS16550 Functional and Timing Considerations[4]

The primary difference between these parts is in the operation of the FIFOs. The NS16550 will sometimes transfer extra characters when the CPU reads the RX FIFO. Due to the asynchronous nature of this failure, there is no work-around, and the NS16550 should *not* be used in the FIFO mode.

The NS16550A/AF has no problems operating in the FIFO modes. The NS16550AF should be used in all new designs.

The programmer should note the difference in the function of Bit 6 in the interrupt identification register (IIR6). This bit is permanently at logic 0 in the NS16550. In the NS16550A/AF, the bit will be set to a 1 when the FIFOs are enabled.

In both parts, Bit 7 of the IIR is set to 1 when the FIFOs are enabled. Therefore, the program can distinguish when the FIFOs are enabled and whether the part is an NS16550A/AF or an NS16550 by checking these two bits.

In order to enable the FIFO mode and set IIR6 and IIR7, Bit 0 of the FIFO control register (FCR0) should be set. Unless both bits are set, the program should not transfer data via the FIFOs.

Bench-Top Operation of the NS16550A/AF UART

Figure 7.14 is the author's bench test setup for these devices. You will notice that the TR1602 is a part of this operation. The two work together

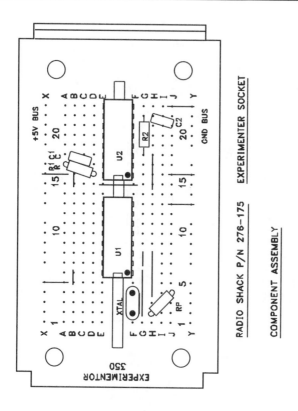

RADIO SHACK P/N 276-175 EXPERIMENTER SOCKET

COMPONENT ASSEMBLY

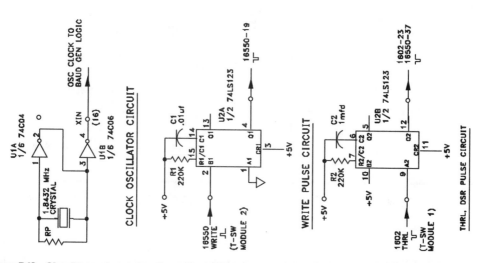

Figure 7.12. Circuitry and prototype board layout for the 1.843-MHx clock and the dual monostable for the NS16550A/AF data bus write (/WR) and 1602 Transmitter Holding Function Load (THRL) functions.

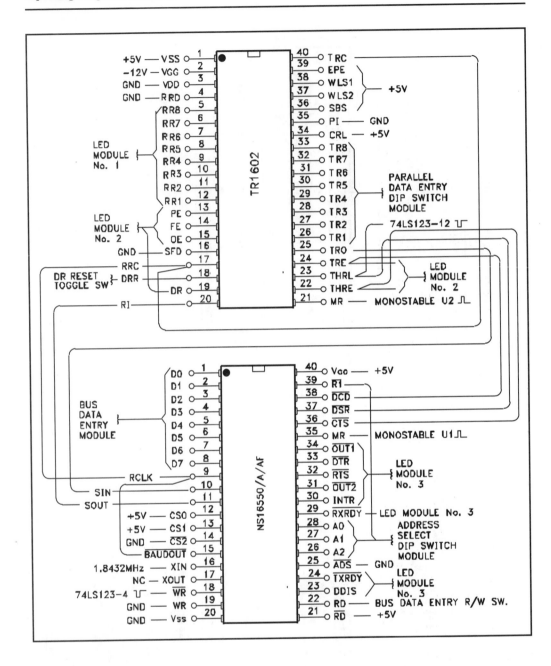

Figure 7.13. An interconnecting planning diagram for linking the TR1602 with the NS16550A/AF. You should use this diagram along with Figure 7.14 when wiring the operating configuration.

very well. The planning diagram of *Figure 7.13* provides a connection overview that will be helpful in the wiring of your own setup.

Previous chapters in this book have provided suggestions for the construction of the modules and their corrugated cardboard mountings, so that will not be repeated here. Do take care with the connections, double-checking as you go, as with the congestion it is easy to poke a wire into an incorrect point.

If you have already developed your test configuration for the devices of Chapter 6, then you are in luck—only one new wire is required: connecting pin 29 (RXRDY) to the end LED of Module 3. The 8250/16450 CSOUT to DS7 of this module is now TXRDY.

If you have bypassed Chapter 6, review the switch usage described in Table 7.9, as there is a departure from previous practice with these devices. Entering write commands is not linked to the clock here on the bench.

Also, I found these will not respond to the high-to-low transition of the bounceless switches employed with the modules. For that reason, you will find a monostable circuit for this purpose, as shown in *Figure 7.12*. The sharp negative-going pulse of the 74LS123 does the job very nicely. The other half of the mono loads the TR1602 transmitter-holding register (THRL). Again, this is identical to that of Chapter 6.

The 1.8432-MHz clock circuitry is also shown with this drawing, as is the layout on an experimenter socket. It didn't seem worthwhile to make a permanent module of these.

I chose a baud generator divisor of 7200 to provide a sufficiently low frequency to observe transfer operations with the LED indicators for SIN and SOUT. Individual bits will still go by too quickly to observe in detail, but the operation will be observable. Decimal 7200 is hexadecimal 1C20 = binary 0001 1100 0010 0000.

The conversion arithmetic is 7200 = 4096 + 2048 + 1024 + 32. The /BAUDOUT frequency is 256 Hz, which is divided by 16 internally to yield 16 Hz. If you have an oscilloscope or frequency counter, they are helpful in assuring the correct frequency but not essential. If you are careful in entering the data on the bus, it will come out correctly.

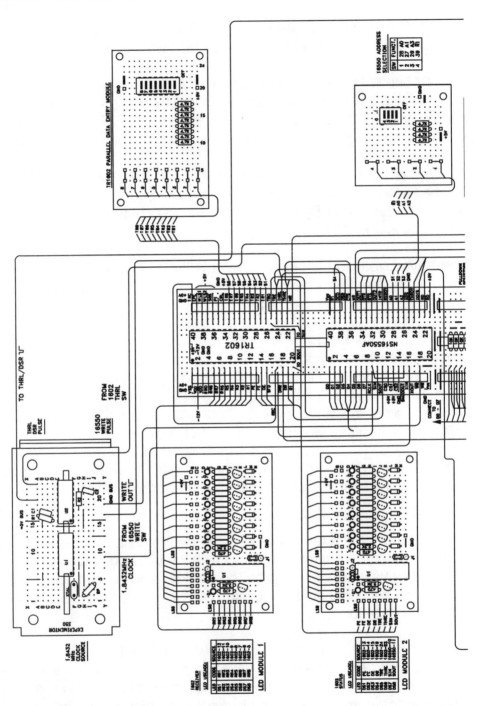

Figure 7.14. The author's bench-top module layout and and wiring configuration for operation of the NS16550A/AF UART. *(continued on next page.)*

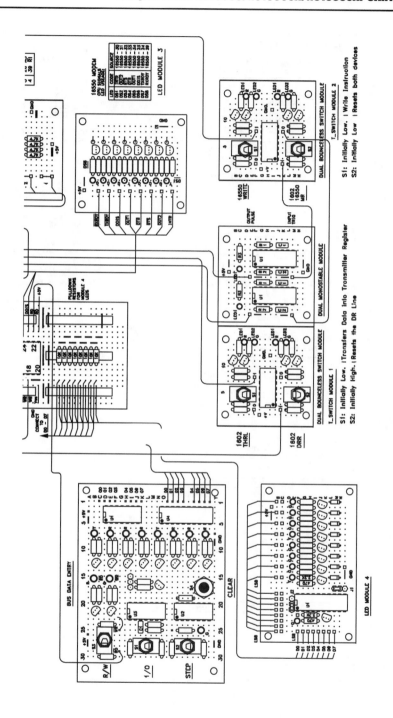

```
INITIALIZING THE NS16550/A/AF UARTNS16450 MODE

CAUTION: Review Table 7.9 before beginning this procedure
to ensure correct R/W switch and DIP switch useage.
NOTE: The LEDs in Modules 1 and 2 may show random
displays on power up.
1. Access the Line Control Register and  ☐
   enable DLAB for setting the Baud Rate
   Generator Divisor Latches.
      The required Baud Rate Divisor is Decimal 7200
                             Binary 0001 1100 0010 0000
```

```
    OFF
1 A0 [▪  ] 1       | 7 6 5 4 3 2 1 0 | OPERATION
2 A1 [▪  ] 1       | 1 0 0 0 0 0 0 0 | Access the DLAB
3 A2 [  ▪] 0
4 -- [ ▪ ] -  ☐    BUS DATA ENTRY = 1000 0000 ☐
      1  0
   REG ADDRESS
   DIP SW SETTINGS
```

```
2. Set the Lower Baud Rate Generator Latch
   to the Least Significant Baud Rate Divsior  ☐

    OFF
1 A0 [▪  ] 0       | 7 6 5 4 3 2 1 0 | OPERATION
2 A1 [  ▪] 0       | 0 0 1 0 0 0 0 0 | LSB Entry
3 A2 [  ▪] 0
4 -- [ ▪ ] -  ☐    BUS DATA ENTRY = 00100 0000 ☐
      1  0
   REG ADDRESS
   DIP SW SETTINGS
```

```
3. Set the Upper Baud Rate Generator Latch
   to the Most Significant Baud Rate Divisor  ☐

    OFF
1 A0 [▪  ] 1       | 7 6 5 4 3 2 1 0 | OPERATION
2 A1 [  ▪] 0       | 0 0 0 1 1 1 0 0 | MSB Entry
3 A2 [  ▪] 0
4 -- [ ▪ ] -  ☐    BUS DATA ENTRY = 0001 1100 ☐
      1  0          Observe a pulse train at pin 15 (BAUDOUT)
   REG ADDRESS      of 256 Hz ☐
   DIP SW SETTINGS
```

```
4. Access the Line Control Register to format
   the Asynchronous Data Communications protocol.  ☐

    OFF
1 A0 [▪  ] 1       | 7 6 5 4 3 2 1 0 | OPERATION
2 A1 [▪  ] 1       | 0 0 1 1 1 0 1 1 | Data Communications
3 A2 [  ▪] 0
4 -- [ ▪ ] -  ☐    BUS DATA ENTRY = 0011 1011 ☐
      1  0          Bit 5: Stick Parity
   REG ADDRESS      Bit 4: Select Even Parity
   DIP SW SETTINGS  Bit 3: Enable Parity
                    Bits 1,0: Word Length = 8 Bits
```

```
5. Access the Interrupt Enable Register.  ☐

    OFF
1 A0 [▪  ] 1       | 7 6 5 4 3 2 1 0 | OPERATION
2 A1 [  ▪] 0       | 0 0 0 0 1 1 1 1 | Interrupts enabled
3 A2 [  ▪] 0
4 -- [ ▪ ] -  ☐    BUS DATA ENTRY = 0000 1111 ☐
      1  0          Intr LED ON ☐
   REG ADDRESS
   DIP SW SETTINGS
```

```
16550
INITIAL STATUS OF LED MODULE 3

DS1 INTR  - OFF
DS2 OUT1  - ON
DS3 DTR   - ON
DS4 RTS   - ON
DS5 OUT2  - ON
DS6 DDIS  - OFF in Read,
          - ON in Write
DS7 TXRDY - OFF
DS8 RXRDY   OFF


1802/16550
INITIAL STATUS OF LED MODULE 2

DS1 PE   - OFF*
DS2 FE   - OFF*
DS3 OE   - OFF*
DS4 DR   - OFF
DS5 TRE  - ON
DS6 THRE - ON
DS7 SIN  - ON (16550)
DS8 SOUT - ON (16550)
* May be ON

1802
INITIAL STATUS OF LED MODULE 1

All dark or random

6. Proceed to the Loop, Run
   or FIFO procedure.
```

Figure 7.15. The Initialization procedure for the NS16550A/AF UART.

```
NS16550/A/AF DATA COMMUNICATION WITH THE TR1602 UART
NS16450 MODE
CAUTION: Review Table 7.9 before beginning this procedure
  to ensure correct R/W switch and DIP switch useage.
The Initialization procedure must have been performed.

1. Setup the TR1602 for Data Communications.

   OFF
   1 TR1  [■□]  1      Optional Data Entry pattern for the
   2 TR2  [□■]  0      TR1602 Parallel Data DIP switch □
   3 TR3  [■□]  1
   4 TR4  [■□]  1
   5 TR5  [■□]  1
   6 TR6  [□■]  0
   7 TR7  [□■]  0
   8 TR8  [■□]  1
        1  0

2. Enable the Data Communications via the MODEM Control Register. □

   OFF
   1 A0  [■□]  0        7 6 5 4 3 2 1 0  | OPERATION
   2 A1  [■□]  0        0 0 0 0 1 1 1 1  | Normal operation mode
   3 A2  [□■]  1
   4 --  [□■]  -  □     BUS DATA ENTRY = 0000 1111 □
        1  0           Observe status of LED Module 3 indicators:
   REG ADDRESS          RXRDY - ON
   DIP SW SETTINGS      OUT1  - OFF
                        DTR   - OFF
                        RTS   - OFF
                        OUT2  - OFF
                        INTR  - ON    □

3. Enter the Data for Transmission on the Bus. □

   OFF
   1 A0  [□■]  0        7 6 5 4 3 2 1 0  | OPERATION
   2 A1  [□■]  0        1 1 0 1 1 0 1 1  | Optional Pattern
   3 A2  [□■]  0
   4 --  [□■]  -  □     BUS DATA ENTRY = 1101 1011 □
        1  0           LED Module 2, DSB blinks with data      □
   REG ADDRESS          LED Module 1 LEDs display data          □
   DIP SW SETTINGS      Toggle DRR                              □
                        Toggle THRL                             □
                        LED Module 2, DS7 blinks with data      □
                        The READ display now matches the TR1602 Data □
                        The Interrupt LED is On                 □

4. Read the MODEM Status Register. □

   OFF
   1 A0  [■□]  0        7 6 5 4 3 2 1 0  | OPERATION
   2 A1  [□■]  1        0 0 0 0 1 0 1 1  | MODEM Status
   3 A2  [□■]  1
   4 --  [■□]  -  □     BUS DATA ENTRY = Read Only    □
        1  0
   REG ADDRESS
   DIP SW SETTINGS

5. Read the Line Status Register. □

   OFF
   1 A0  [■□]  1        7 6 5 4 3 2 1 0  | OPERATION
   2 A1  [□■]  0        0 1 1 0 0 0 0 0  | Line Status
   3 A2  [■□]  1
   4 --  [■□]  -  □     BUS DATA ENTRY = Read Only □
        1  0           Bit 5: THRE
   REG ADDRESS          Bit 6: TEMT
   DIP SW SETTINGS      Bit 1, 2, or 3 On is error indication
```

Figure 7.16. The Run procedure for the NS16550A/AF UART.

NS16550/A/AF DATA DIAGNOSTIC LOOPING PROCEDURE NS16450 MODE

CAUTION: Review Table 7.9 before beginning this procedure
to ensure correct R/W̄ switch and DIP switch useage.
The Initialization procedure must have been performed.

1. Enable the Internal Diagnostic Loop via the MODEM Control Register. ☐

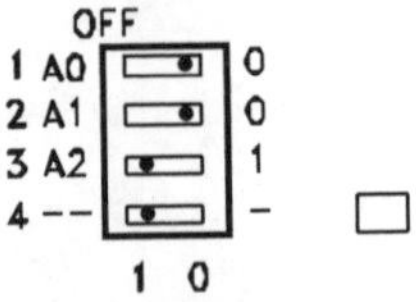

OFF

1 A0	▭	0
2 A1	▭	0
3 A2	▭	1
4 --	▭	- ☐

1 0

REG ADDRESS
DIP SW SETTINGS

7 6 5 4 3 2 1 0	OPERATION
0 0 0 1 0 0 0 0	Loop Operation

BUS DATA ENTRY = 0001 0000 ☐

Observe status of LED Module 3 indicators:

RXRDY – ON
OUT1 – ON
DTR – ON
RTS – ON
OUT2 – ON
INTR – ON ☐

2. Enter the Data for Transmission on the Bus ☐
   Perform a write, followed by the Read

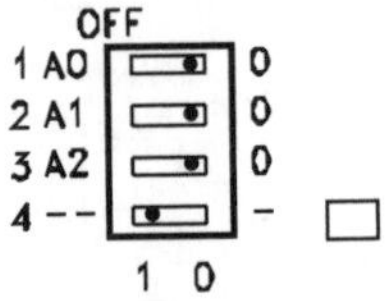

OFF

1 A0	▭	0
2 A1	▭	0
3 A2	▭	0
4 --	▭	- ☐

1 0

REG ADDRESS
DIP SW SETTINGS

7 6 5 4 3 2 1 0	OPERATION
1 1 0 1 1 0 1 1	Optional Pattern

BUS DATA ENTRY = 1101 1011 ☐
The Bus Read should be identical to the Write.

3. Read the MODEM Status Register ☐

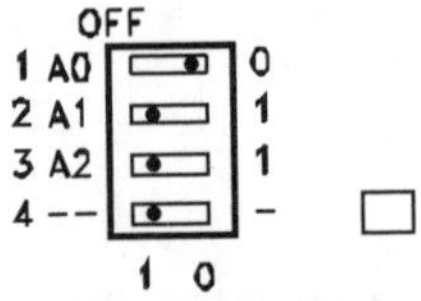

OFF

1 A0	▭	0
2 A1	▭	1
3 A2	▭	1
4 --	▭	- ☐

1 0

REG ADDRESS
DIP SW SETTINGS

7 6 5 4 3 2 1 0	OPERATION
0 0 0 0 1 0 1 1	MODEM Status

BUS DATA ENTRY = Read Only ☐
☐

4. Read the Line Status Register ☐

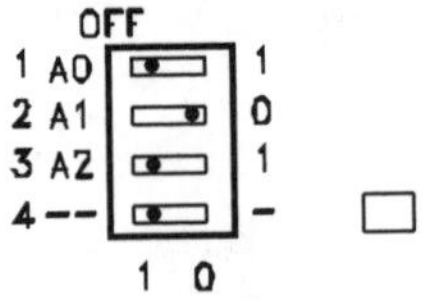

OFF

1 A0	▭	1
2 A1	▭	0
3 A2	▭	1
4 --	▭	- ☐

1 0

REG ADDRESS
DIP SW SETTINGS

7 6 5 4 3 2 1 0	OPERATION
0 1 1 0 0 0 0 0	Line Status

BUS DATA ENTRY = Read Only ☐
☐
Bit 5: THRE
Bit 6: TEMT

Figure 7.17. The Loop procedure for the NS16550A/AF UART.

There are four test procedures for these devices detailed in *Figures 7.15, 7.16, 7.17,* and *7.18.* By no means do these procedures explore all the possibilities of these remarkable devices. But they provide us with insights into their operations and hopefully will inspire your own further explorations beyond that provided.

Tables 7.10, 7.11, and *7.12* are summaries of the control and status register functions. I found these helpful for ready reference when working up the test procedures. The first procedure, Figure 7.15, provides the device initialization routine. As mentioned previously we access the line control register (LCR) at 011 and set Bit 7 to logic 1 for access to the divisor latches. After entering the latch values we return to the LCR to set the protocols and exit the divisor latches. The procedure sets the data word at eight bits. You may want to experiment with other values.

The next procedure can be any of the remaining three—run, FIFO, or loop—whichever you desire to perform next. These can be performed in any sequence without having to reinitialize. If you perform a reset, however, then steps 4 and 5 of Figure 7.15 will need reentering before proceeding.

The procedures as given perform the data communication between the NS16550A/AF and the TR1602, though observing interrupt operation will not be all that could be hoped for. This is an area where experimenting can be rewarding. The handshaking is embedded with the TR1602 interface; if desired, you can replace these functions with another four-position DIP switch module and enter them yourself.

It will be informative to change data values for repeat transfers to better see how these transactions take place. If this is the first device you have selected for bench operation it will take a little experience to enter the data correctly. Just be patient, do a restart and continue.

Each procedure is designed to be self-explanatory. Even so, it will be helpful to review the text on related features and the procedures with care before and/or during your operation of the device. I have found these to be challenging devices to master. But I have also found satisfaction in the achievement. I'm sure you will also.

NS16550/A/AF DATA COMMUNICATION WITH THE TR1602 UART

NS16550A MODE FIFO ENABLED

CAUTION: Review Table 7.9 before beginning this procedure to ensure correct R/W switch and DIP switch usage. The Initialization procedure must have been performed.

1. Setup the TR1602 for Data Communications.

OFF

1	TR1	1	Optional Data Entry pattern for the
2	TR2	0	TR1602 Parallel Data DIP switch ☐
3	TR3	1	
4	TR4	1	
5	TR5	1	
6	TR6	0	
7	TR7	0	
8	TR8	1	

1 0

2. Enable the FIFO Register and define the trigger level. ☐

OFF

1	A0	0
2	A1	1
3	A2	0
4	--	-

1 0

REG ADDRESS
DIP SW SETTINGS

7 6 5 4 3 2 1 0	OPERATION
1 0 0 0 1 0 0 1	FIFO Mode

BUS DATA ENTRY = 1000 1001 ☐

Trigger set for 8 bytes

NOTE: The FIFO Register is Write only. With the R/W switch in the Read position it is the Interrupt Identification Register that is read.

RXRDY LED ON ☐
INTR LED ON ☐

Interrupt Identification Register Read

7 6 5 4 3 2 1 0	OPERATION
1 1 0 0 0 0 1 0	FIFO Mode

Status: FCR0 = 1. No interrupt pending

3. Enable the Data Communications via the MODEM Control Register. ☐

OFF

1	A0	0
2	A1	0
3	A2	1
4	--	-

1 0

REG ADDRESS
DIP SW SETTINGS

7 6 5 4 3 2 1 0	OPERATION
0 0 0 0 1 1 1 1	Normal operation mode

BUS DATA ENTRY = 0000 1111 ☐

Observe status of LED Module 3 indicators:

RXRDY — ON
TXRDY — OFF
OUT1 — OFF
DTR — OFF
RTS — OFF
OUT2 — OFF
INTR — ON ☐

Figure 7.18. A FIFO procedure for the NS16550A/AF UART. *(Continued on next page.)*

4. Enter the Data for Transmission on the Bus ☐
   Write one or more entries, optional patterns
   Return the R/W to Read

```
OFF
1 A0 [ ▪ ] 0
2 A1 [ ▪ ] 0
3 A2 [ ▪ ] 0
4 -- [ ▪ ] -    ☐
    1 0
REG ADDRESS
DIP SW SETTINGS
```

7 6 5 4 3 2 1 0	OPERATION
1 1 0 1 1 0 1 1	Optional Pattern

BUS DATA ENTRY = 1101 1011 ☐
LED Module 2, DS8 blinks with data   ☐
LED Module 1 LEDs display data   ☐
Toggle DRR   ☐
Toggle THRL   ☐
LED Module 2, DS7 blinks with data   ☐
The READ display now matches the TR1602 Data ☐
The Interrupt LED is On   ☐
Possible PE, OE, FE indication, LED Mod. 2   ☐

5. Read the MODEM Status Register. ☐

```
OFF
1 A0 [ ▪ ] 0
2 A1 [ ▪ ] 1
3 A2 [ ▪ ] 1
4 -- [ ▪ ] -    ☐
    1 0
REG ADDRESS
DIP SW SETTINGS
```

7 6 5 4 3 2 1 0	OPERATION
0 0 0 0 1 0 1 1	MODEM Status

BUS DATA ENTRY = Read Only   ☐

6. Read the Line Status Register. ☐

```
OFF
1 A0 [ ▪ ] 1
2 A1 [ ▪ ] 0
3 A2 [ ▪ ] 1
4 -- [ ▪ ] -    ☐
    1  0
REG ADDRESS
DIP SW SETTINGS
```

7 6 5 4 3 2 1 0	OPERATION
0 1 1 0 0 0 0 0	Line Status

BUS DATA ENTRY = Read Only ☐
   Bit 5: THRE
   Bit 6: TEMT
   Bit 1, 2, or 3 On is error indication

7. Reset the FIFO Register to the NS16450 Mode.   ☐
   This is a write only register.
   Return to Read shows Interrupt Priority status.

```
OFF
1 A0 [ ▪ ] 0
2 A1 [ ▪ ] 1
3 A2 [ ▪ ] 0
4 -- [ ▪ ] -    ☐
    1 0
REG ADDRESS
DIP SW SETTINGS
```

7 6 5 4 3 2 1 0	OPERATION
1 0 0 0 0 0 0 0	NS16450 MODE

BUS DATA ENTRY = 1000 0000 ☐
Trigger remains set for 8 bytes ☐
RXRDY – ON ☐

CONTROL REGISTER FUNCTION SUMMARY		
BH	**Register/Function**	**Operation**
	Interrupt Enable Register IER Address: 001	
0	Received Data Available Interrupt Also timeout when in the FIFO mode.	Interrupt Enabled when set to logic 1
1	Transmitter Holding Register Empty Interrupt	Interrupt Enabled when set to logic 1
2	Receiver Line Status Interrupt	Interrupt Enabled when set to logic 1
3	MODEM Status Interrupt	Interrupt Enabled when set to logic 1
4—7	Always 0	
	Line Control Register LCR Address 011	
0	Word Length Select	Refer to Table 6.6
1	Word Length Select	Refer to Table 6.6
2	Specifies number of Stop Bits	0=1, 1=1 1/2 if 5bit data, else 2
3	Parity Enable	1=Parity bit generated
4	Even Parity Select	Bits 3 and 4=1 even parity, 1 and 0, odd parity
5	Stick Parity	1 with BH 3=1 parity detected as 0 if BH 4=1
6	Set Break	1=SOUT forced Low, 0=Disabled
7	DLAB Access	1= Access to Baud Rate Gen Divisor Latches
	MODEM Control Register MCR Address 100	
0	Data Terminal Ready (DTR)	1= DTR forced LOW
1	Request To Send (RTS)	1= RTS forced LOW
2	OUT 1	1= OUT1 forced LOW
3	OUT2	1= OUT2 forced LOW
4	LOOP	1= following procedure: SOUT set to logic 1 SIN disconnected Transmitter Shift Register output looped back into Receiver Shift Register input. MODEM control inputs CTS, DSR, DCD, RI disconnected MODEM control outputs DTR, RTS, OUT1, OUT2 internally connected to the four MODEM control inputs. Transmitted data is immediately received. Receiver and Transmitter Interrupts are operational. MODEM control Interrupts are operational, sources are the four lower bits of the MCR. Interrupts still controlled by the IER.
5—7	Always 0	

Table 7.10.   A summary of control register functions: the Interrupt Enable (IER), the Line Control (LCR), and the MODEM Control (MCR).

Table 7.11. *(At right)* **A summary of the FIFO control register functions.**

FIFO CONTROL REGISTER FUNCTION SUMMARY

FIFO Control Register Address 010

Bit	Register/Function	Operation
0	FIFO Enable	Writing a 1 enables XMIT and RCVR FIFOs
1	RCVR FIFO Reset	Writing a 1 clears RCVR FIFO, resets its counter logic
2	XMIT FIFO Reset	Writing a 1 clears XMIT FIFO, resets its counter logic
3	DMA Mode Select	Writing a 1 causes RXRDY, TXRDY to change mode 0 to mode 1
4	Reserved for future use	
5	Reserved for future use	
6	RCVR Trigger LSB	Writing a 1 sets trigger level RCVR FIFO interrupt, LSB
7	RCVR Trigger MSB	Writing a 1 sets trigger level RCVR FIFO interrupt, MSB

STATUS REGISTER FUNCTION SUMMARY		
Bit	Register/Function	Operation
	Interrupt Identification Register IIR Address 010	
0	Interrupt Status	0 if Interrupt pending
1	Interrupt Identification, bit 0	Establishes priority with bit 2
2	Interrupt Identification, bit 1	Establishes priority with bit 1
3	Set along with Bit 2 when a timeout Interrupt is pending	Establishes priority with bit 2
4,5	Always 0	
6	FIFOs enabled	Set when FCR0=1
7	FIFOs enabled	Set when FCR0=1
	Line Status Register LSR Address 101	
0	Data Ready DR	1= complete incoming char received and is in the Receiver Buffer Register Bit is reset by CPU read or reset input
1	Overrun Error OE	1= Rcvr Buf Data overrun with new data
2	Parity Error PE	1= Received character has incorrect parity
3	Framing Error FE	1= Received character has incorrect stop bit
4	Break Interrupt BI	1= Received character held in 0 state beyond the full word transmission time
5	Transmitter Holding Register Empty THRE	1= on transfer to the Shift Reg, issues interrupt
6	Transmitter Empty TEMT	1= Transmitter Shift Register is idle. Is reset on data transfer from the Holding Register
7	Receiver FIFO error	1= At least one PE, OE, or FE error
	MODEM Status Register MSR Address 110	
0	Data Clear To Send (DCTS)	1= $\overline{CTS}$ input has changed state since CPU read
1	Delta Data Set Ready (DDSR)	1= $\overline{DSR}$ input has changed state since CPU read
2	Trailing Edge Ring Indicator (TERI)	1= indicates RI change, logic 1 to 0
3	Delta Data Carrier Detect (DDCD)	1= $\overline{DCD}$ input has changed state
		NOTE: Whenever an above bit is set, a MODEM Status Interrupt is generated
4	Clear To Send (CTS)	Complements the $\overline{CTS}$ input
5	Data Set Ready (DSR)	Complements the $\overline{DSR}$ input
6	Ring Indicator (RI)	Complements the $\overline{RI}$ input
7	Delta Carrier Detect (DCD)	Complements the $\overline{DCD}$ input

Table 7.12. A summary of status register functions: the Interrupt Identification (IER), the Line Status (LSR), and the Modem Status (MSR).

References

1. National Semiconductor Corporation, *Data Communications Local Area Networks/UARTS,* "NS16550AF Universal Asynchronous Receiver/Transmitter with FIFOs," 1990, pp. 4-36 to 4-56.

2. Ibid, Martin S. Michael and Daniel G. Durich, Application Note 491, "The NS16650A: UART Design and Application Considerations," pp. 4-57 to 4-83.

3. Registered trademark of National Semiconductor Corporation.

4. Ibid, Martin S. Michael, Application Note 493, "A Comparison of the INS8250, NS 16450 and NS16550 Series of UARTs," p. 4-85.

5. Reference for Figure 7.1: Ibid, "Basic Configuration," p. 4-36.

6. Reference for Figure 7.2: Ibid. 5.0 Block diagram, p. 4-45, Connection diagrams, p. 4-48.

7. Reference for Figure 7.3: Ibid, 9.0 Typical Applications, p. 4-55.

8. Reference for Figure 7.4: Ibid. 8.2 Typical Clock Circuits, p. 4-50.

9. Reference for Figure 7.5: Ibid. 3.0 AC Electrical Characteristics, Baud Generator p. 4-39, 4.0 Timing Waveforms, /BAUDOUT Timing, p. 4-40.

10. Reference for Figure 7.6: Ibid. 3.0 AC Electrical Characteristics, p. 4-39, 4.0 Timing Waveforms, Read and Write Timing, p. 4-41.

11. Reference for Figure 7.7: Ibid. 3.0 AC Electrical Characteristics, Receiver p. 4-39, Transmitter p. 4-40, 4.0 Timing Waveforms, Receiver and Transmitter Timing, p. 4-42.

12. Reference for Figure 7.8: Ibid. 3.0 AC Electrical Characteristics, Modem Control p. 4-40, 4.0 Timing Waveforms, Modem Controls Timing, p. 4-42.

13. Reference for Figure 7.9: Ibid. 4.0 Timing Waveforms, pp. 4-43, 4-44.

14. Reference for Figure 7.10: Ibid. 4.0 Timing Waveforms, pp. 4-44.

15. Reference for Figure 7.11: Ibid. Typical Applications, p. 4-56.

16. Reference for Table 7.1: Ibid. 6.0 Pin Descriptions, Table: Register Addresses, p. 4-46.

17. Reference for Table 7.2: Ibid. Table I, UART Reset Configuration, p. 4-48.

18. Reference for Table 7.3: Ibid. Table 7.III, "Baud Rates Using 1.8432 MHz Crystal," and Table 7.IV, "Baud Rates Using 3.072 MHz Crystal," pp. 4-50, 4-51.

19. Reference for Table 7.4: Ibid. Table 7.V, "Baud Rates Using 8 MHz Crystal," p. 4-51.

20. Reference for Table 7.5: Ibid. Table II, "Summary of Registers," p. 4-49.

21. Reference for Table 7.6: Ibid. 8.1 Line Control Register, p. 4-50.

22. Reference for Table 7.7: Ibid. 8.5 FIFO Control Register, p. 4-52.

23. Reference for Table 7.8: Ibid. Table VI, "Interrupt Control Functions," p. 4-52.

Chapter 8
The 8251/8251A USART

Introduction

The 8251/8251A is a universal synchronous/asynchronous receiver/transmitter (USART) capable of operating with a variety of serial communication interfaces[1-3]. The device is a component of the Intel MCS-80[4] microprocessor family and, as such, is capable of interfacing to a variety of microcomputers, including the 8048, 8080, 8085, 8086, 8088, and Z80[5] systems with a minimum of additional hardware.

The 8251/8251A can support most serial data techniques, including the IBM "bi-sync." *Figure 8.1* illustrates a typical microcomputer application. The addressing shown is for device (chip) selection (CS) and control vs. data input (C//D).

In the receive mode, the device converts incoming serial formatted data into parallel format and makes certain checks on its correct transmission. In the transmit mode, parallel data is converted to serial format for sending. It also inserts and deletes characters and bits unique to the type of transmission.

Table 8.1 summarizes the 8251/8251A major features. We see that the device is extremely versatile, with two operating modes and a high degree of capability in its performance. The 8251A is an advanced design of the 8251. *Table 8.2* is a listing of the additional features and enhancements. Many of these are safeguards as well as an improvement in overall performance.

Figure 8.2 shows the DIP packaging, with the 28 pin functions identified, a description of the pin functions, and a block diagram illustrating the internal operations.

Table 8.3 describes the AC characteristics for both the 8251 and the enhanced 8251A.

Figures 8.3 through *8.7* are timing diagrams. These should be reviewed with some care—the chip's internal operation is critical in its timing—particularly the distinctions between the asynchronous and synchronous modes.

The next group of figures relate to the device operation. We must gain familiarity with these for our operation on the bench.

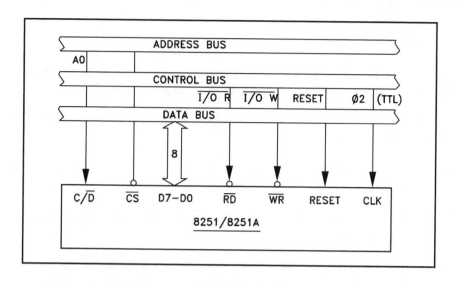

Figure 8.1. The 8251/8251A interface to the 8080 Standard System Bus.

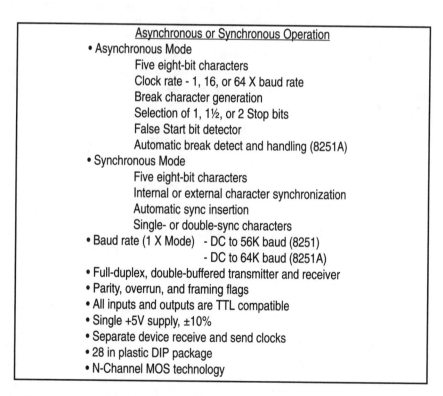

<u>Asynchronous or Synchronous Operation</u>
- Asynchronous Mode
  - Five eight-bit characters
  - Clock rate - 1, 16, or 64 X baud rate
  - Break character generation
  - Selection of 1, 1½, or 2 Stop bits
  - False Start bit detector
  - Automatic break detect and handling (8251A)
- Synchronous Mode
  - Five eight-bit characters
  - Internal or external character synchronization
  - Automatic sync insertion
  - Single- or double-sync characters
- Baud rate (1 X Mode) - DC to 56K baud (8251)
  - DC to 64K baud (8251A)
- Full-duplex, double-buffered transmitter and receiver
- Parity, overrun, and framing flags
- All inputs and outputs are TTL compatible
- Single +5V supply, ±10%
- Separate device receive and send clocks
- 28 in plastic DIP package
- N-Channel MOS technology

Table 8.1. Significant features of the 8251/8251A USART.

The 8251A USART vs. The 8251 USART

1. The data are double-buffered with separate I/O registers for control, status, Data In and Data Out. This feature simplifies control programming and minimizes processor overhead.

2. The Receiver detects and handles "break" automatically in asynchronous operations, which relieves the processor of this task.

3. The Receiver is prevented from starting when in "break" state by a refined Rx initialization. This also prevents a disconnected USART from causing unwanted interrupts.

4. When a transmission is concluded, the TxD line will always return to the marking state unless SBRK is programmed.

5. The Tx Disable command is prevented from halting transmission by the Tx Enable Logic enhancement, until all data previously written has been transmitted. The same logic also prevents the transmitter from turning off in the middle of a word.

6. Internal Sync Detect is disabled when External Sync Detect is programmed. An External Sync Detect is provided through a flip-flop, which clears itself upon a status read.

7. The possibility of false sync detect is minimized by:
   - ensuring that if a double sync character is programmed, the characters be contigously detected.
   - clearing the Rx register to all logic 1s (VOH) whenever the Enter Hunt command is issued in Sync Mode.

8. The /RD and /WR do not affect the internal operation of the device as long as the 8251A is not selected.

9. The 8251A Status can be read at any time; however, the status update will be inhibited during a status read.

10. The 8251A has enhanced AC and DC characteristics and is free from extraneous glitches, providing higher speed and improved operating margins.

11. Baud rate from DC to 64K.

Table 8.2. Features and enhancements of the 8251A over the 8251.

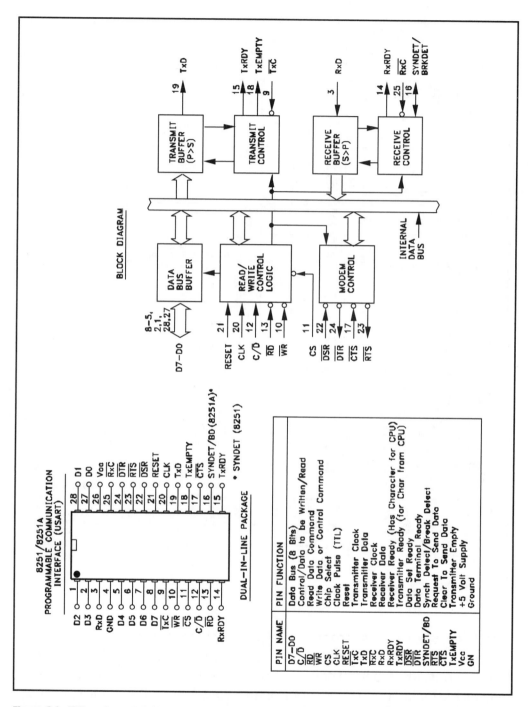

Figure 8.2. DIP package details with the pin functions described and the internal block diagram for the 8251/8251A USART.

8251/8251A AC CHARACTERISTICS							
		LIMITS					
		8251		8251A			TEST
PARAMETER	SYMBOL	MIN	MAX	MIN	MAX	UNIT	CONDITIOIN
READ							
Address Strobe before READ, (CS,C/D)	tAR	50		0		nS	
Address Hold Time for READ, (CS,C/D)	tRA	5		0		nS	
READ Pulse Width	tRR	430		250		nS	
Data Delay from READ	tRD		350		200	nS	8251 CL=100pf
							8251A CL=150pf
READ to Data Floating	tDF		200		100	nS	8251 CL=100pf
		25		10			CL=15pf
WRITE							
Address Strobe before WRITE	tAW	20		0		nS	
Address Hold Time for WRITE	tWA	20		0		nS	
WRITE Pulse Width	tWW	400		250		nS	
Data Setup Time for WRITE	tDW	200		150		nS	
Data Hold Time for WRITE	tWD	40		0		nS	
Recovery Time between WRITES ②	tRV	6		6		tCY	
OTHER TIMING							
Clock Perios ③	tCY	0.420	1.35	0.32	1.35	uS	
Clock Pulse Width High	tØW	220	0.7CY	120	tCY−90	nS	
Clock Pulse Width	tØW			90		nS	
Clock Rise and Fall Time	tR,tf	0	50	5	20	nS	
TxD Delay from Falling Edge of TxC	tDTx		1		1	uS	8251 CL=100pf
Rx Data Setup Time to Sampline Pulse	tSRx	2		2		uS	8251 CL=100pf
Rx Data Hold Time to Sampling Pulse	tHRx					uS	8251 CL=100pf
Transmitter Input Clock Frequency	tTX						
1X Baud Rate		DC	56		64	KHz	
16X Baud Rate		DC	520		310	KHz	
64X Baud Rate		DC	520		615	KHz	
Transmitter Input Clock Pulse Width	tTPW						
1X Baud Rate		12		12		tCY	
16X and 64X Baud Rate		1		1		tCY	
Transmitter Input Clock Pulse Delay	tTPD						
1X Baud Rate		15		15		tCY	
16X and 64X Baud Rate		3		3		tCY	
Receiver Input Clock Frequency	tRX						
1X Baud Rate		DC	56		64	KHz	
16x Baud Rate		DC	520		310	KHz	
64X Baud Rate		DC	520		615	KHz	
Receiver Input Clock Pulse Width	tRPW						
1X Baud Rate		12		12		tCY	
16X and 64X Baud Rate		1		1		tCY	
Receiver Input Clock Delay	tRPD						
1X Baud Rate		15		15		tCY	
16X and 64X Baud Rate		3		3		tCY	
TxRDY Delay from Center of Data Bit	tTx		16		8	tCY	8251 CL=50pf
RxRDY Delay from Center of Data Bit	tRx		20		24	tCY	
Internal SYNDET Delay from Center of Data Bit	tIS		25		24	tCY	
External SYNDET Setup Time before Falling Edge of RxC	tES		18		16	tCY	
TxEMPTY Delay from Center of Data Bit	tTxE		16		20	tCY	8251 CL=50pf
Control Delay from Rising Edge of WRITE (TxE,DTR,RTS)	tWC		16		8	tCY	
Control to READ Setup Time (DSR,CTS)	tCR		18		20	tCY	

NOTES: 1. AC Timing measured at VOH =2.0, VOL=0.8 and with the load circuit shown
2. This recovery time is for initialization only when MODE, SYNC1, SYNC2 COMMAAND and first DATA BYTES are written into the USART. Subsequent writing of both COMMAND and DATA are only allowed when TxRDY=1.
3. The TxC and RxC frequencies have the following limitations with respect to the Clock:
For 1X Baud Rate, fTx or fRx < 1/(30 tCY)
For 16X and 64X Baud Rate, fTx or fRx < 1/(4.5 tCY)
4. Reset Pulse Width = 6 tCY mimimum

Table 8.3. The 8251 and 8251A AC characteristics.

The 8251/8251A differ from the INS8250/NS16450/NS16550 series we studied in Chapters 6 and 7 in that data entry is not clocked into or out of the data bus. Internally, of course, clocking is a requirement. There are two distinct operating conditions: control and data. Within the control, we have mode selection and command instructions. As seen in *Figure 8.8*, the active

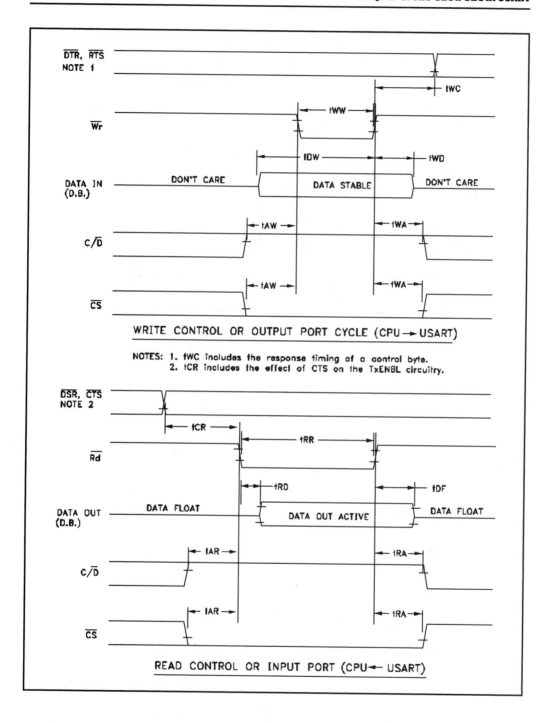

NOTES: 1. tWC includes the response timing of a control byte.
2. tCR includes the effect of CTS on the TxENBL circuitry.

Figure 8.3. Timing diagrams for Write and Read control of the 8251/8251A USART.

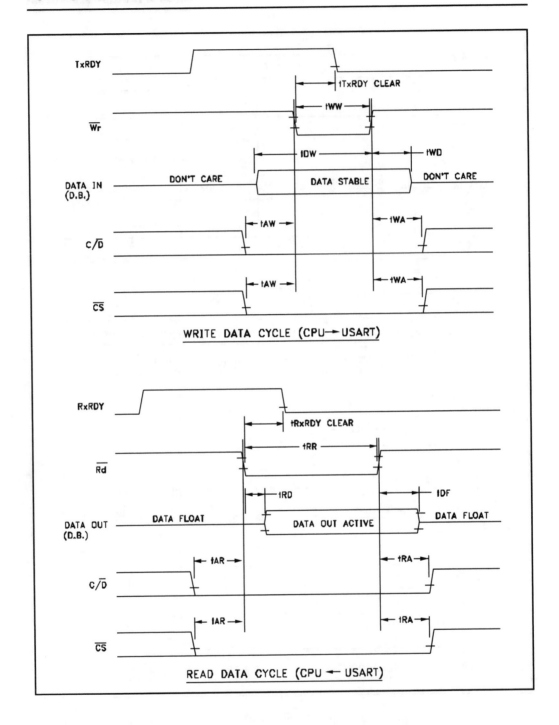

Figure 8.4. Write and Read Data Cycle timing diagrams of the 8251/8251A USART.

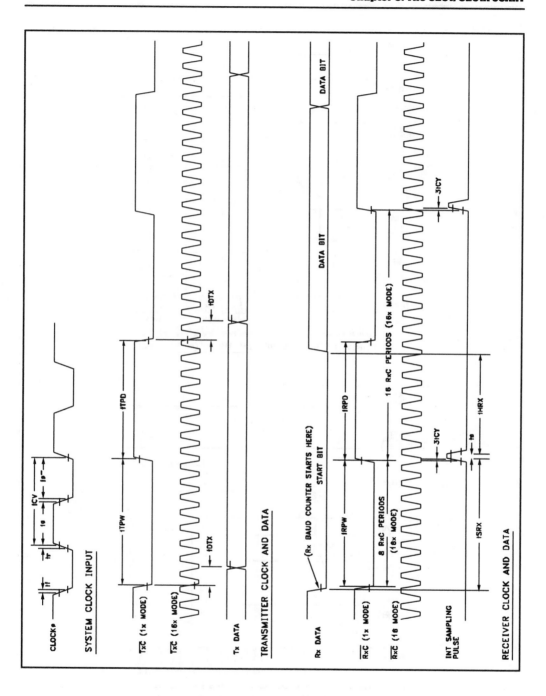

Figure 8.5. Transmitter and Receiver Clock and Data timing diagrams of the 8251/8251A USART.

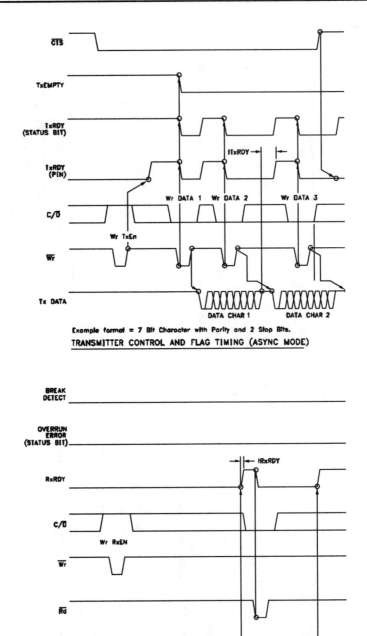

Example format = 7 Bit Character with Parity and 2 Stop Bits.
TRANSMITTER CONTROL AND FLAG TIMING (ASYNC MODE)

Example format = 7 bit character with parity and two stop bits.
RECEIVER CONTROL AND FLAG TIMING (ASYNC MODE)

Figure 8.6. Transmitter and Receiver Control and Flag Timing diagrams for the asynchronous mode of the 8251/8251A USART. *(Continued on next page.)*

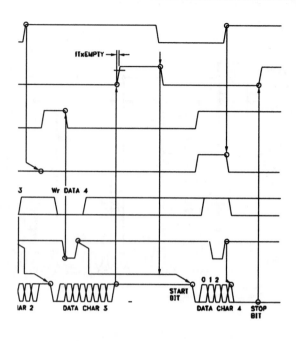

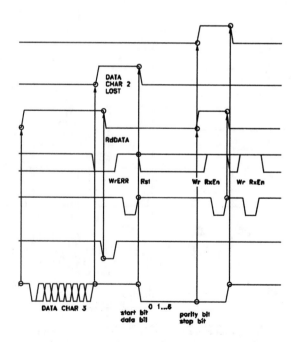

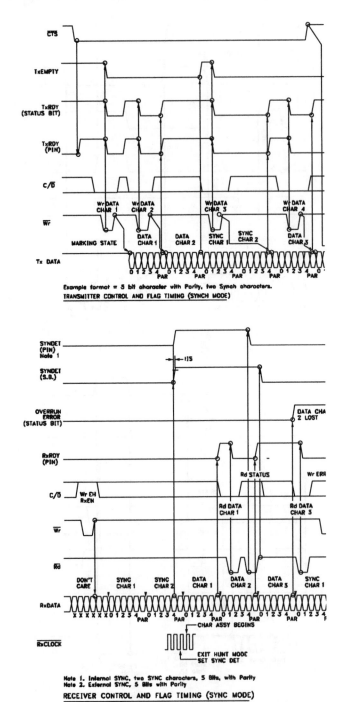

Figure 8.7. Transmitter and Receiver Control and Flag Timing diagrams for the synchronous mode of the 8251/8251A USART. *(Continued on next page.)*

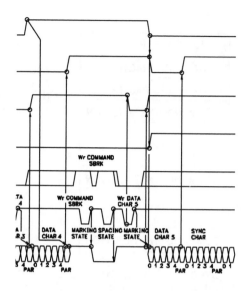

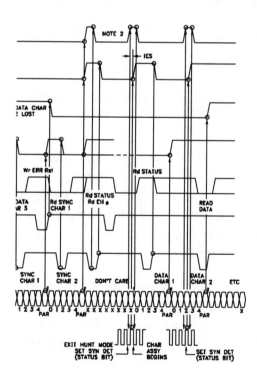

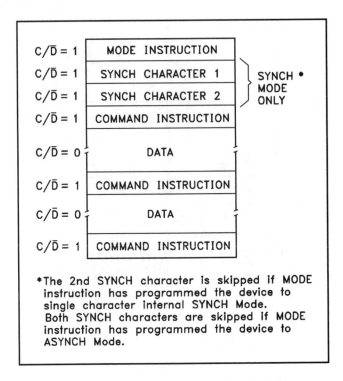

Figure 8.8. A typical block of control for the 8251/8251A USART illustrating the status of the C//D input and the sequence of Mode, Command, and Data entries to the bus.

condition is determined by the status of input C//D. With this input high (logic "1"), entries to the data bus are accepted as controls. With this input low (logic "0"), entries are accepted as data. Typically, the mode instruction is a one-time requirement. The asynchronous/synchronous mode selection is made with this entry. Command instructions that follow may be revised as operational needs determine.

Figure 8.9 defines the instruction format for the asynchronous mode. As we shall see, the format for this mode differs greatly from that for the synchronous mode. In this figure, each of the eight bits in the instruction is defined. Thus, if we wish an eight-bit data word, then we set bits 0 and 1 to logic "1." If we wish the synchronous mode, then we set these two to logic "0" and move on to *Figure 8.10.* In addition to the instruction formats, these two figures illustrate the structure of the serial data word.

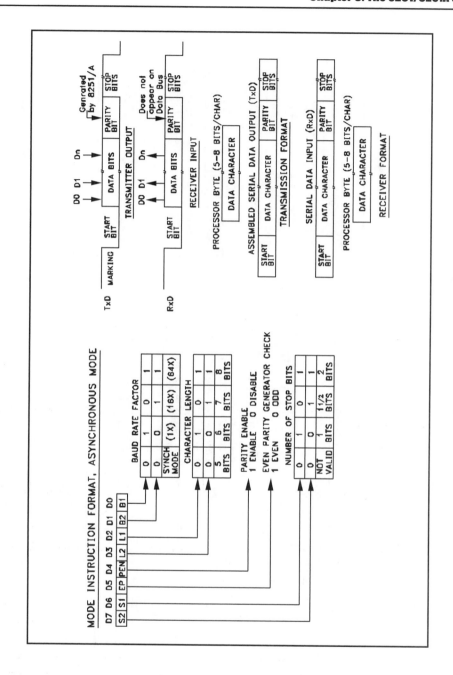

Figure 8.9. The Asynchronous Mode Instruction format for the 8251/8251A USART.

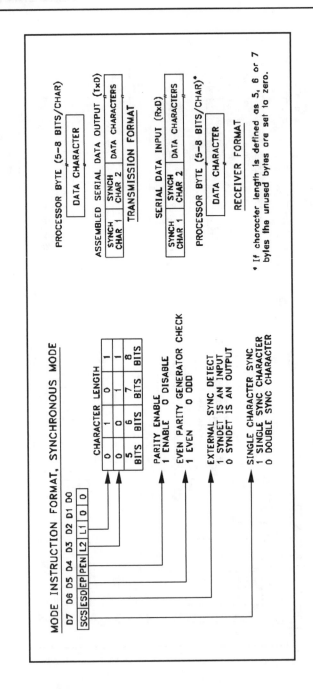

Figure 8.10. The Synchronous Mode Instruction format for the 8251/8251A USART.

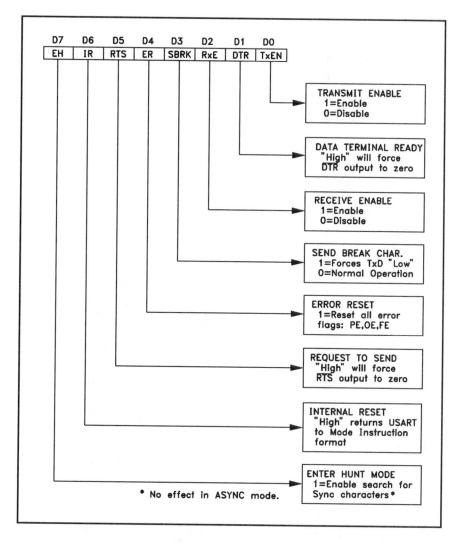

Figure 8.11. The Control Instruction format for the 8251/8251A USART. Note that this format applies to both the asynchronous and synchronous modes.

Figures 8.9 and 8.10 showed the format for the asynchronous and synchronous modes, respectively. *Figure 8.11* provides the command instruction format for both modes. We see that the only difference is with bit 7, which has application to the synchronous mode only.

When the command mode bus is in the read state, the data read describes the system status. It is referred to as the status read. Thus, the CPU can

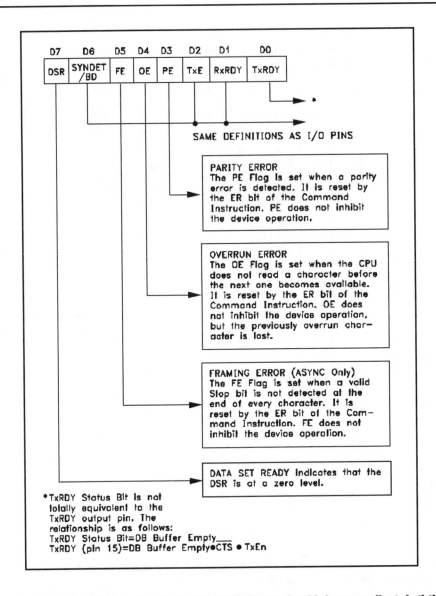

Figure 8.12. The Status Read format for the 8251/8251A USART. Note that this format applies to both the asynchronous and synchronous modes.

query the device for its internal status at any time. *Figure 8.12* describes the bus format for the status read.

The remaining figures relate to the bench-top operation and are described in that section of the chapter.

USART Pin Descriptions

The descriptions that follow are organized by function blocks. The function acronym is followed by the pin number(s) and a summary of the function. Note that in this book, "/" defines the complement—that is, the term is active low.

Function	Pin(s)	Function
D7-D0	1,2,27 28, 5-8	Data Bus Buffer. An eight-bit, three-state bidirectional buffer interfacing the USART to the CPU data bus. Data is transmitted or received by the buffer in response to input/output or read/write instructions from the processor. Control words, command words, and status are also transferred on the bus.
Vcc	26	+5V power supply.
GND	4	Ground.

Read/Write Control Logic

This logic block accepts inputs from the processor control bus and generates signals for overall USART operation. The mode-instruction and command-instruction registers that store the control formats for device functional definition are located in this logic block.

Function	Pin(s)	Function
RESET	21	A logic "1" on this input forces the USART into the "idle" mode, where it will remain until reinitialized with a new set of control words. Minimum RESET pulse width is 6tCY.
CLK	20	Clock Pulse. The CLK input provides for internal device timing and is usually connected to the Phase 2 (TTL) output of the 8224 clock generator. External inputs and outputs are not referenced to CLK, but the CLK frequency must be at least 30 times the receiver or transmitter clocks in the synchronous mode and 4.5 times in the asynchronous mode.
/WR	10	Write Data. A "0" on this active low input instructs the USART to accept the data or control word, which the processor is writing out on the data bus.

/RD	13	Read Data. A "0" on this active low input instructs the USART to place the data or status information onto the data bus for the processor to read.
C//D	112	Control/Data. This input, in conjunction with the /WR and /RD inputs, informs the USART to accept or provide either a data character, control word, or status information via the data bus. 0=Data, 1=Control.
/CS	11	Chip Select. A "0" on this input enables the USART to read from or write to the processor.

Modem Control

The USART includes a set of control inputs and outputs that may be used to simplify the interface to a modem.

Function	Pin(s)	Function
/DSR	22	Data Set Ready. This input can be tested by the processor via status information. The /DSR input is normally used to test modem data set ready condition.
/DTR	24	Data Terminal Ready. This output can be controlled via the command word. The /DTR output is normally used to drive modem data terminal ready or rate-select lines.
/RTS	23	Request-to-Send. This output can be controlled via the command word. The /RTS output is normally used to drive the modem request-to-send line.
/CTS	17	Clear-to-Send. A "0" on this input enables the USART to transmit serial data if the TxEN bit in the command-instruction register is enabled (a logic 1).

Transmit Buffer

The transmit buffer receives parallel data from the data-bus buffer via the internal data bus, performs a parallel-to-serial conversion on the data, inserts the necessary characters or bits needed for the programmed communication format, and outputs composite serial data on the TxD pin. The transmit control logic accepts and outputs all external and internal signals necessary for serial data transmission.

Function	Pin(s)	Function
TxRDY	15	Transmitter Read. This output informs the processor that the transmitter is ready to accept a data character. TxRDY may be used as an interrupt or may be tested through the status information for polled operation. Loading a character from the processor automatically resets TxRDY on the leading edge.
TxE	18	Transmitter Empty. This output signals the processor that the USART has no further characters to transmit. TxE is automatically reset upon receiving a data character from the processor. In half-duplex, TxE can be used to signal end of a transmission and request the processor to "turn the line around." The TxEN bit in the command instruction does not effect TxE. In the synchronous mode, a "1" on this output indicates that the a sync character or characters are about to be automatically transmitted as "fillers," because the next data character has not been loaded.
/TxC	9	Transmitter Clock. This clock controls the serial character-transmission rate. In the asynchronous mode, the /TxC frequency is a multiple of the actual baud rate. Two bits of the mode instruction select the multiple to be 1X, 16X, or 64X the baud rate. In the synchronous mode, the /TxC frequency is automatically selected to equal the actual baud rate. Note that for both synchronous and asynchronous modes, serial data is shifted out by the falling edge of /TxC.
TxD	19	Transmitter Data. The transmit control logic outputs the composite serial data stream on this pin.

Receive Buffer

The receive buffer accepts serial data input at the /RxD pin and converts the data from serial-to-parallel format. Bits or characters required for the specific communication technique in use are checked and then an eight-bit "assembled" character is readied for the processor. For communication techniques that require less than eight bits, the USART sets the extra bits to zero. The receiver control logic manages all activities related to incoming data.

Function	Pin(s)	Function
RxRDY	14	Receiver Ready. This output indicates that the receiver buffer is ready with an "assembled" character for input to the processor. For polled operation, the processor can check RxRDY using a status read, or RxRDY can be con-

nected to the processor interrupt structure. Note that reading the character to the processor automatically resets RxRDY.

/RxC	25	Receiver Clock. This clock determines the rate at which the incoming character is received. In the asynchronous mode, the /RxC frequency may be 1X, 16X, or 64X the actual baud rate, but in the synchronous mode, the /RxC frequency must equal the baud rate. Two bits in the mode instruction select asynchronous at 1X, 16X, or 64X, or select synchronous at 1X the baud rate. Unlike /TxC, data is sampled by the USART on the rising edge of /RxC. Note: Since the USART will frequently be handling both the reception and transmission for a given link, the receive and transmit baud rates will be the same. /RxC and /TxC then require the same frequency and may be tied together and connected to a single clock source or the baud-rate generator. Example: If the baud rate equals 110 (async): /RxC or /TxC equals 110 Hz (1X) /RxC or /TxC equals 1.76 kHz (16X) /RxC or /TxC equals 7.04 kHz (64X) If the baud rate equals 300: /RxC or /TxC equals 300 Hz (1X) A or S /RxC or /TxC equals 4800 Hz (16X) A only /RxC or /TxC equals 19.2 kHz (64X) A only
RxD	3	Receiver Data. A composite serial data stream is received by the receiver control logic on this pin.
SYNDET (8251)	16	Sync Detect. This pin function is only used in the synchronous mode. The 8251 may be programmed through the mode instruction to operate in either the internal or external sync mode and SYNDET, then functions as an output or input, respectively. In the internal sync mode, the SYNDET output will go to a "1" when the 8251 has located the sync character in the receive mode. If double sync character (bi-sync) has been programmed, SYNDET will go to "1" in the middle of the last bit of the second sync character. SYNDET is automatically reset to "0" upon a status read or reset. In the external sync mode, a "0" to "1" transition on the SYNDET input will cause USART to start assembling data characters on the next falling edge of /RxC. The length of the SYNDET input should be at least one /RxC period but may be removed once the USART is in sync.
SYNDET/BD (8251A)	16	Sync Detect/Break Detect. This pin is used in both the synchronous and asynchronous modes. When in the sync mode, the features for the SYNDET pin as previously described apply. When in asynchronous mode, the break detect output will go high when an all-zero word of the programmed length is received. This word consists of start bit, data bit, parity bit, and one stop bit. Reset only occurs when the Rx data returns to a logic 1 state or upon chip reset. This state of break detect can be read as a status bit.

Operating the 8251/8251A USART

The text that follows refers to both the 8251 and the 8251A unless otherwise indicated.

Operational Start-Up Requirements

An initial set of control words must be sent to the device via the data bus to define the desired mode (synchronous or asynchronous) and the communications format. The control words must specify the baud-rate factor (1x, 16x, 64x), the character length (five to eight), the number of stop bits (1, 1½, 2), the mode (asynchronous or synchronous), SYNDET (in or out), and parity. Once the control words have been entered, the device is ready to communicate. TxRDY is raised to a logic "1" to signal the processor that the USART is ready to receive a character for transmission. When the processor writes a character to the USART, TxRDY is automatically reset.

The USART may receive serial data concurrently with serial transmission. After the receipt of a completed character, the RxRDY output is raised to indicate that a completed character is ready for the processor. The processor fetch will automatically reset the RxRDY.

CAUTION: The device may provide faulty RxRDY for the first read after power-up, since the first read received after power-up is enabled by the command instruction (RxE). A dummy read is recommended to clear a possible faulty RxRDY. This will not be an occurrence for the first read after a hardware or software reset once the operation has been established.

Data transmission cannot take place until the TxEN bit has been set by a command instruction and until the /CTS (clear-to-send) input is a logic "0." TxD is held in the "marking" state after reset awaiting new control words.

USART Programming

The USART must be loaded with a group of two to four control words provided by the processor before data communication can begin. A reset (internal or external) must immediately follow the control words, which have been entered to program the complete operational interface for data communication. Note that if an external reset is not available, three successive 000Hex or two successive 80Hex instructions (C/D=1)—followed by a software reset command instruction (40Hex)—can be used to initialize the device. Keep in mind that there are two control word formats: mode instruction and command instruction.

Mode Instruction

This control word specifies the general characteristics of the interface regarding the synchronous or asynchronous mode, baud-rate factor, character length, parity, and number of stop bits. Once the mode instruction has been received, SYNC characters or command instructions may be inserted, depending on the mode instruction content.

Command Instruction

This control word will be interpreted as a SNYC character definition, if immediately preceded by a mode instruction that specified a synchronous format. After the SYNC character(s) are specified, or after an asynchronous mode instruction, all subsequent control words will be interpreted as an update to the command instruction. Command instruction updates may occur at any time during the data block. To modify the mode instruction, a bit may be set in the command instruction that causes an internal reset, allowing a new mode instruction to be accepted. Figure 8.8 illustrates the sequence of a typical entry mix of mode instruction, command instruction, and data.

Mode Instruction Definition

Understanding how the mode instruction controls the functional operation of the device is easiest when the USART is viewed as two separate component structures—one asynchronous, the other synchronous—that share the same support circuitry and package. The format can be changed as needed or "on the fly," so to speak.

Asynchronous Transmission

When data is written into the USART, the device automatically inserts the start bit (low level or space) and the number of stop bits (high level or mark), as specified by the mode instruction. If parity has been enabled, an odd or even parity bit is inserted just before the stop bit(s). Then, depending on /CTS and TxEN, the character may be transmitted as a serial data stream at the TxD output. Data is shifted out by the falling edge of /TxC at /TxC, /TxC/16 or /TxC/64, as defined by the mode instruction.

Asynchronous Receive

The RxD input line is normally held high (marking) by the transmitting device. A falling edge at RxD signals the possible beginning of a start bit and a new character. The start bit is checked by testing for a "low" at its nominal center, as specified by the baud rate. If a low is detected again, it is considered valid, and the counter employed for assembling the bit begins its operation. The bit counter locates the approximate center of the data, parity (if specified), and stop bits.

The parity error flag (PE) is set in the event that a parity error has occurred. Input bits are sampled at the RxD pin with the rising edge of /RxC. If a high is not detected for the stop bit, which normally signals the end of an input character, a framing error flag (FE) is set. After a valid stop bit, the input character is loaded into the parallel data buffer and the RxRDY signal is raised to indicate to the processor that a character is ready to be fetched.

If the processor has failed to fetch the previous character, the new character replaces the old and the overrun error flag (OE) is set. All the error flags can be reset by setting a bit in the command instruction. Error flag conditions will not stop subsequent USART operations.

Figure 8.9 illustrates the asynchronous mode instruction format and the communications format for the transmission and receive operations and data structures.

Synchronous Transmission

As in asynchronous transmission, the TxD output remains high (marking) until the device receives the first (usually a SYNC) character from the processor. After a command instruction has set TxEN, and after clear-to-send (/CTS) has gone low, the first character is serially transmitted. Data is shifted out on the falling edge of /TxC at the same rate as /TxC.

Once transmission has started, synchronous mode format requires that the serial data stream at TxD continue at the /TxC rate, or else SYNC will be lost. If a data character is not provided by the processor before the device's transmit buffer becomes empty, the SYNC character(s) loaded directly following the mode instruction will be automatically inserted into the TxD data stream. The SYNC characters are inserted to fill the line and maintain synchronization until new data characters are available for transmission.

If the USART does become empty and must send SYNC characters, the TxEMPTY output is raised to signal the processor that the transmitter buffer is empty and SYNC characters are being transmitted. TxEMPTY is automatically reset by the next character from the processor.

Synchronous Receive

In synchronous receive, the character synchronization can be either external or internal. If the internal SYNC mode has been selected and the enter HUNT (EH) bit has been set by a command instruction, the receiver goes into the HUNT mode.

Incoming data on the RxD input is sampled on the rising edge of /RxC, and the receive buffer is compared with the first SYNC character after each bit has been loaded until a match is found. If two SYNC characters have been programmed, the next received character is also compared. When the SYNC character(s) programmed have been detected, the USART leaves the HUNT mode and is in character synchronization. At this time, the SYNDET (output) is set high. SYNDET is automatically reset by a status read.

If external SYNC has been specified in the mode instruction, a "1" applied to the SYNDET (input) for at least one /RxC cycle will synchronize the USART.

Parity and overrun errors are treated the same in the synchronous mode as in the asynchronous. If not in HUNT, parity will continue to be checked even if the receiver is not enabled. Framing errors do not apply in the synchronous format. The processor may command the receiver to enter the HUNT mode with a command instruction, which sets enter HUNT (EH) if synchronization is lost.

Figure 8.10 illustrates the synchronous mode instruction format and the communications format for the transmission and receive operations and data structures.

Command Instruction Format

After the functional definition of the USART has been specified by the mode instruction and the SYNC characters have been entered (if in SYNC mode), the device is ready to receive the command instructions and begin communication. A command instruction is used to control the specific operation of the format selected by the mode instruction. Enable transmit, enable receive, error reset, and modem controls are controlled by the command instruction.

After the mode instruction and the SYNC character(s) (as needed) are loaded, all subsequent "control writes" (C//D=1) will load or overwrite the command instruction register.

A reset operation (internal via CMD IR or external via the RESET input) will cause the USART to interpret the next "control write"—which must immediately follow the reset—as a mode instruction.

Figure 8.11 illustrates the command instruction format with a definition of each bit provided.

Status Read Format

It is frequently necessary for the processor to examine the status of an active interface device to determine if errors have occurred or if there are other conditions that require a response from the processor. The USART has features that allow the processor to read the device status at any time.

A data fetch is issued by the processor while holding the C//D input high to obtain device status information. Many of the bits in the status register are copies of external pins. This dual status arrangement allows the device to be used in both polled and interrupt-driven environments. Status updates can have a maximum delay of 16 clock periods in the 8251 and 28 clock periods in the 8251A. Figure 8.12 illustrates the status read format with a definition of each bit provided.

Parity Error

When a parity error is detected, the PE flag is set. The flag is cleared by setting the ER bit in a subsequent command instruction. PE being set doesn't inhibit USART operation.

Overrun Error

If the processor fails to read a data character before the one following is available, the OE flag is set. The flag is cleared by setting the ER bit in a subsequent command instruction. Although OE being set doesn't inhibit USART operation, the previously written character has been written over and is lost.

Framing Error

If a valid stop bit is not detected at the end of a character, the FE flag is set. The flag is cleared by setting the ER bit in a subsequent command instruction. FE being set doesn't inhibit USART operation. Framing error detection is a function provided in the asynchronous mode only.

Bench-Top Operation of the 8251/8251A USART

The 8251/8251A USART is unique with respect to the previous programmable devices in that it features two independent modes of operation: synchronous and asynchronous. Operation in the asynchronous mode can be performed in combination with the TR1602 UART, as was done in Chapters 6 and 7.

But this approach is not available to us with the synchronous mode, so we need to find an alternative method. If you are not familiar with the TR1602, it would be helpful to review the description in Chapter 3 prior to performing the procedure for the asynchronous mode.

Another complexity is the allowable range of the CLK input, which is restricted to a period of 0.42 µS to 1.35 µS and to a frequency range of 740 Hz to 2,381,000 Hz. In the asynchronous mode, the frequency of CLK must be greater than 4.5 times the baud rate input and 30 times greater when in the synchronous mode. But it's desirable to maintain a low frequency for the baud rate to obtain a visible indication of the data transfer. Thus, two widely divergent clock sources are needed. It is our good fortune that data bus read and write and the external inputs are independent of the clocks.

Figure 8.13 describes the circuitry and prototype construction for a 1.8432-MHz CLK source. This is the same clock employed with the devices of Chapters 6 and 7.

A second circuit on the assembly is a dual monostable. One section of this is triggered by the output of another clock source, providing the baud-rate frequency. This circuitry and its construction are shown in *Figure 8.14.* The other half of the monostable is triggered by the TR1602 THRL toggle switch to provide the pulse for this TR1602 function.

The baud-rate frequency is derived from a 7555 CMOS timer oscillator running at a nominal 4800 Hz. Two CD4017 decade counter/dividers reduce

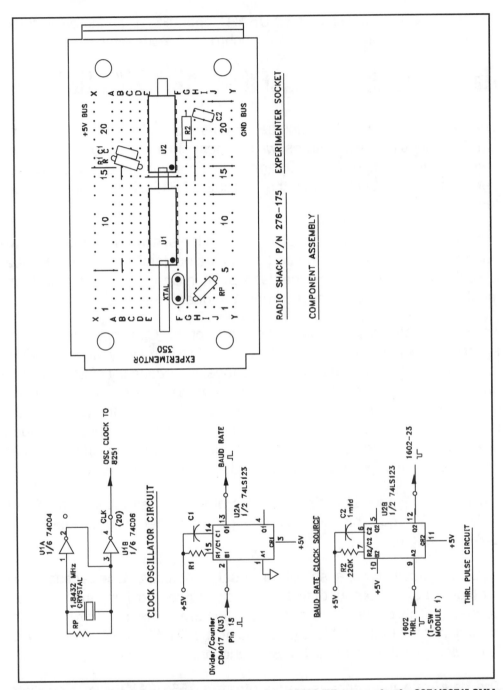

Figure 8.13. The circuitry and prototype assembly for (a) a 1.8432-MHz source for the 8251/8251A CLK input, and (b) a dual monostable: one circuit for the baud-rate input for the 8251/8251A and the TR1602, the other as the pulse source for the TR1602 THRL input.

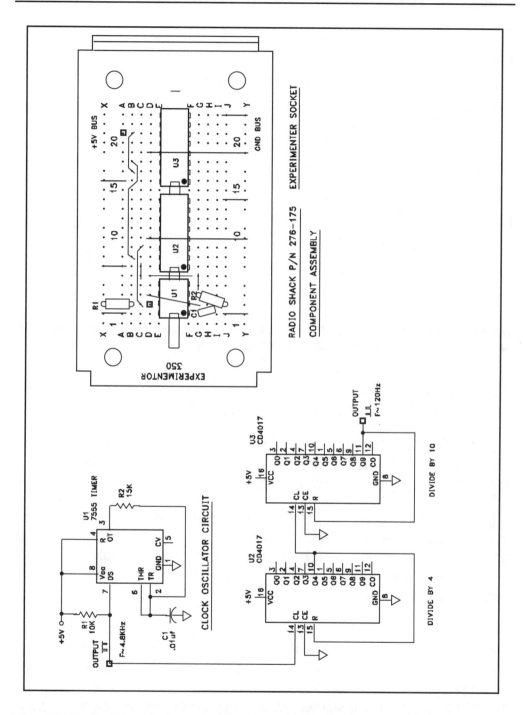

Figure 8.14. The circuitry and prototype assembly for a 7555 CMOS clock and counter/dividers functioning as the trigger source for the baud-rate monostable circuit of Figure 8.13.

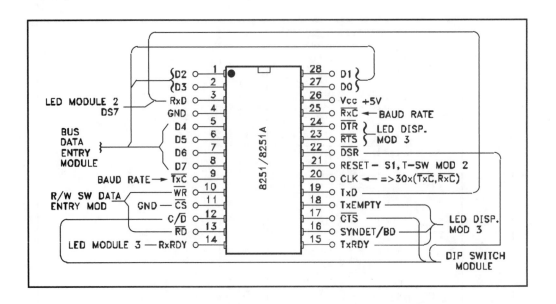

Figure 8.15. The connection-planning diagram for the 8251/8251A USART in the synchronous communication mode. You should use this diagram with Figure 8.17 when wiring the operating configuration.

this to approximately 120 Hz. The output pulse is extremely narrow and is used to trigger a monostable for increased stability.

If an oscilloscope is available, it will be helpful in verifying operation of the clocks. An advantage of the low baud rate is the visible indication of serial data transmission on the LED indicators.

Figure 8.15 is a planning diagram for interfacing the 8251/8251A to operate in the synchronous communication mode. It would be nice to have a companion device such as the TR1602, but since none is available the solution is to connect the serial transmission of data from the output pin to the pin for receiving serial input data.

The planning diagram is helpful in wiring the operational configuration of *Figure 8.17*. I can hear someone saying, "Whoa now, this is for the asynchronous mode." True enough, but it also accommodates the synchronous, so simply ignore the TR1602 connections. To gain the greatest understanding of the device operation, you should run it in both modes anyway. *Figure 8.16* is the planning diagram for inclusion of the TR1602.

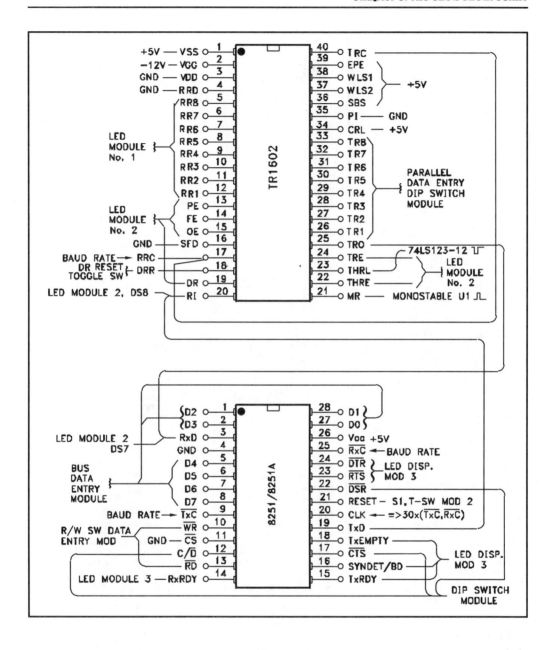

Figure 8.16. The connection planning diagram for the 8251/8251A USART in the asynchronous communication mode. You should use this diagram with Figure 8.17 when wiring the operating configuration.

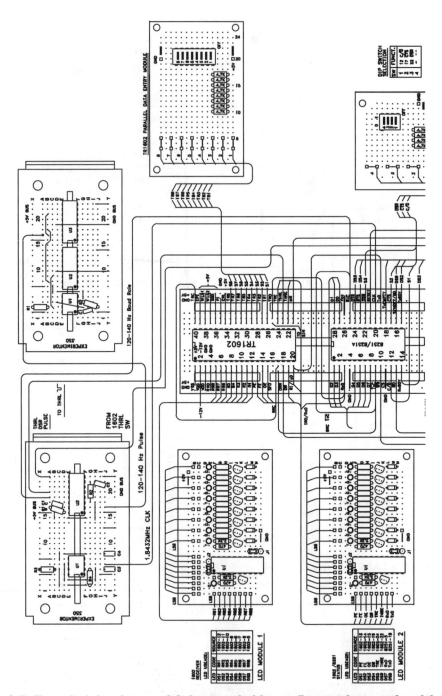

Figure 8.17. The author's bench-top module layout and wiring configuraton for operation of the 8251/8251A USART. The TR1602 UART is employed with the asynchronous communication mode only. *(Continued on next page.)*

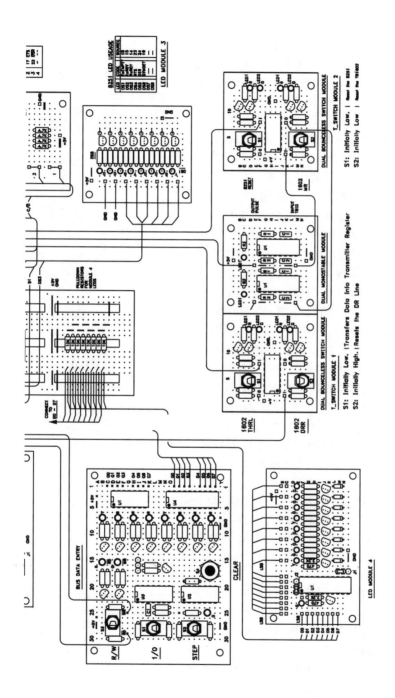

The SYNDET function is used only in the synchronous mode. It may be either an input or output. I have chosen to use it as an output. The following will show why.

In the synchronous mode, the receiver clocks in the specified number of data bits and transfers them to the receiver buffer register, setting RxRDY. There must be a means for synchronizing the receiver to the transmitter so that the required character boundaries are maintained. This is met by the use of the sync characters shown on the flowchart. Synchronization is obtained in the HUNT mode, wherein the 8251 shifts in data on the RxD line one bit at a time. With each bit, the register is compared with the register containing a sync character. Shifting continues until the comparison produces a match. This terminates the HUNT mode and raises the SYNDET line.

When used as an input, the synchronization is done externally, which requires timing signals difficult for us to provide when working on the bench.

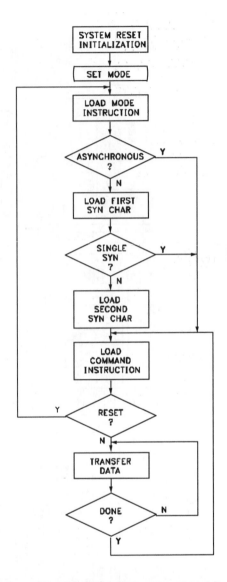

Figure 8.18. The flowchart for operation of the 8251/8251A USART in the synchronous and asynchronous communications modes.

Previous chapters of this book have provided suggestions for the construction of the modules and their corrugated cardboard supports, so these will not be repeated here. Do take care with the connections, double-checking as you go, as with the congested layout it is easy to poke a wire into an incorrect point. If you have already set up your wiring for the devices of Chapters 6 or 7, you will find much of the work already done.

OPERATION OF THE 8251/8251A USART IN THE SYNCHRONOUS MODE

NOTE: Use the R/W switch on the Data Entry Module for Bus Read and Write. Following a Write return the switch to the Read position.

On Power-Up, enter 0000 0000 three times, followed by 0000 0100 for a simulated software reset.

POWER UP STATUS OF LED MODULES

INITIAL STATUS OF LED MODULE 1
Dark: Not used.

1. Initialize the 8251/8251A for operation in the Synchronous Mode. ☐

BUS DATA ENTRY = 1111 1110 ☐

7 6 5 4 3 2 1 0	OPERATION
1 1 1 1 1 1 0 0	Mode Instruction

Synchronous Mode
8-Bit Data Word
Even Parity Enable
Single Sync Char
Internal Sync

OFF
1 C/D [ ] 1
2 CTS [ ] 1
3 DSR [ ] 1
4 -- [ ] - ☐
1   0
DIP SW SETTINGS

8251/8251A
INITIAL STATUS OF LED MODULE 2
DS1 – DS6 Not used.
DS7 RxD – ON
DS8 TxD – ON

7 6 5 4 3 2 1 0	OPERATION
1 0 0 0 0 1 0 1	Status in Read Position

DSR*   TxE TxRDY
If reset

8251/8251A
INITIAL STATUS OF LED MODULE 3
DS1 TxEMPTY – ON
DS2 TxRDY   – OFF
DS3 RxRDY   – OFF
DS4 RTS   – ON
DS5 DTR   – ON
DS6 SYNDET – OFF
DS7 --
DS8 --

2. Enter the Sync Character.   ☐

OFF
1 C/D [ ] 1
2 CTS [ ] 1
3 DSR [ ] 1
4 -- [ ] - ☐
1   0
DIP SW SETTINGS

7 6 5 4 3 2 1 0	OPERATION
0 0 0 1 1 0 1 1	Data Word to Bus

BUS DATA ENTRY = 0001 1011 ☐
Data Bus returns to Status Read as above.

3. Enter a Command Instruction for Send and Receive Operation. ☐

BUS DATA ENTRY = 0010 0111 ☐

OFF
1 C/D [ ] 1
2 CTS [ ] 1
3 DSR [ ] 1
4 -- [ ] - ☐
1   0
DIP SW SETTINGS

7 6 5 4 3 2 1 0	OPERATION
1 0 1 0 0 1 1 1	Command Instruction

Transmit Enable
Set DTR Low
Receive Enable
Normal Operation
Do Not Reset Error Flags
Set RTS Low
Do Not Reset
Enable search for Sync Chars

NOTE: Observe that the DTR and RTS LEDs of Module 3 go Low.
If they do not, repeat the Write Instruction. (Essential)

4. Set C/D to 0 to enter the Data Mode ☐
Set DSR to 0 (Essential)
Return the R/W toggle to the Read position.
Set CTS to 0.
LED Module 2 DS7 respond with a fast blink.
Observe the data on LED Module 4.
Restore CTS to 1 before writing a new data character.

OFF
1 C/D [ ] 0
2 CTS [ ] 1
3 DSR [ ] 1
4 -- [ ] - ☐
1   0
DIP SW SETTINGS

7 6 5 4 3 2 1 0	OPERATION
1 1 0 1 1 0 1 1	Data Word to Bus

BUS DATA ENTRY = 1101 1011 ☐
The data shown is suggested only. Repeat the entry procedure with new data to observe the operation. Note that on return to the Read position the data does not change until CTS is zeroed.
Toggle DSR to observe effect on Status Read.

Figure 8.19. The bench-top initialization and operating procedure for the 8251/8251A USART in the synchronous communications mode.

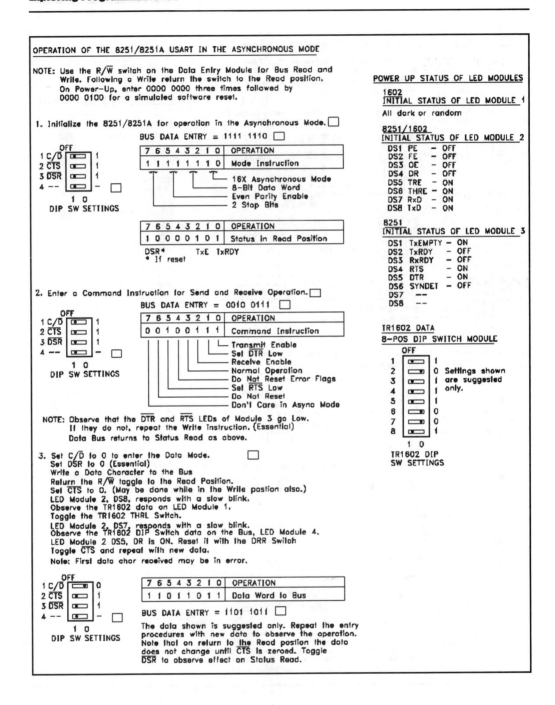

Figure 8.20. The bench-top initialization and operating procedure for the 8251/8251A USART in the asynchronous communications mode.

This is not the easiest device to use in a bench-top operation. It can be fussy, and patience may be required. It is imperative to carry out the initial software reset of three 0000 0000 inputs to the data bus followed by the 0000 0100 input, as called for on the procedure guides. The read/write switch on the data-entry module is used for reading from and writing to the data bus. Because the switch outputs are toggled between high and low states, one line will always be high with the other low.

There are two operating procedures—*Figures 8.19* and *8.20*—one for each of the two communication modes. The procedures are designed to be self-explanatory as much as possible. Even so, it is advisable to review the text before proceeding with their operation.

I did find this a challenging device to master. It can be a frustrating experience. But I have found that perseverance pays dividends in satisfaction. I'm sure that the same will be true for you.

References

1. Intel Corporation, *Microsystem Components Handbook, Vol. 2,* Data Sheet, "8251A Universal Synchronous/Asynchronous Receiver/Transmitter," 1984, p. 7-155.

2. Ibid, pp. 7-3 to 7-31, Lionel Smith, AP-16 "Using the 8251 Universal Synchronous/Asynchronous Receiver Transmitter."

3. NEC Microcomputers, Inc., *Programmable Communication Interfaces,* Data Sheet, "NEC µPD8251/µPD8251A Universal Synchronous/Asynchronous Receiver/Transmitter, Rev/4, p. 583.

4. MCS is a registered trademark of Intel Corporation.

5. Z80 is a registered trademark of Zilog.

6. Reference for Figure 8.1: Intel Corporation, *Microsystem Components Handbook, Vol. 2,* Data Sheet, "8251A Universal Synchronous/Asynchronous Receiver/Transmitter," 1984, Figure 6, p. 7-160.

7. Reference for Figure 8.2: Ibid. Figure 2, p. 7-155.

8. Reference for Figure 8.3: Ibid. Waveforms, "Write Control...," "Read Control...," p. 7-170.

9. Reference for Figure 8.4: Ibid. Waveforms, "Write Data Cycle," "Read Data Cycle," p. 7-169.

10. Reference for Figure 8.5: Ibid. Waveforms, "System Clock Input," "Transmitter Clock and Data," "Receiver Clock and Data," p. 7-169.

11. Reference for Figure 8.6: Ibid. Waveforms, "Receiver Control and Flag Timing (ASYNC MODE)," "Transmitter Control and Flag Timing (ASYNC MODE)," pp. 7-169, 7-170.

12. Reference for Figure 8.7: Ibid. Waveforms, "Receiver Control and Flag Timing (SYNC MODE)," "Transmitter Control and Flag Timing (SYNC MODE)," p. 7-170.

13. Reference for Figure 8.8: Ibid. Figure 7, p. 7-160.

14. Reference for Figure 8.9: Ibid. Figures 7, 8, p. 7-161, 162.

15. Reference for Figure 8.10: Ibid. Figures 10, 11, p. 7-163

16. Reference for Figure 8.11: Ibid. Figure 12, p. 7-164.

17. Reference for Figure 8.12: Ibid. Figure 13, p. 7-164.

18. Reference for Table 8.3: Ibid. AC Characteristics, pp. 7-166, 7-167.

Chapter 9
The 8253/8253-5
Programmable Interval Timer

Introduction

The 8253[1] is a programmable interval timer/counter specifically designed for use with Intel microcomputer systems. Its function is that of a general purpose, multitiming element that can be treated as an array of I/O ports in the system software.

In its system application, the 8253 generates accurate time delays under software control. The programmer configures the 8253 to match the requirements, initializing one of the counters with the required quantity. Upon command, the device will count out the delay and interrupt the CPU when its purpose has been accomplished. Multiple delays can be realized by an assignment of priorities.

Other counter/timer functions that are nondelay common to most microcomputers can be implemented with the 8253. These include:

- Programmable rate generator
- Event counter
- Binary rate multiplier
- Real-time clock
- Digital one-shot
- Complex motor controller

The 8253 System Interface

Figure 9.1 illustrates the device interface in a microcomputer system. The 8253 is a component of Intel microcomputer systems and interfaces in the same manner as other components of the system. The system software treats it as an array of peripheral I/O ports; three are counters, with a fourth control register for mode programming.

The select inputs A0,A1 connect to the A0,A1 address bus signals of the CPU. The /CS can be derived directly from the address bus using a linear select method, or, alternatively, to the output of a decoder, such as the Intel 8205[2].

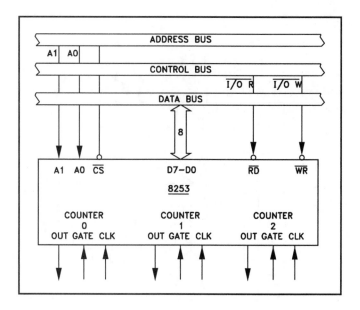

Figure 9.1. The 8253 System Interface.

Figure 9.2 defines the requirement for clock-rate division, with the 8253-5 clocked from the 8085 microprocessor.

Functional Description

Figure 9.3 describes the package pinning, the pin functions, and provides a block diagram of the internal configuration.

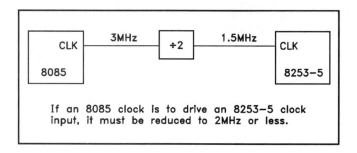

Figure 9.2. 8253-5 clock constraint.

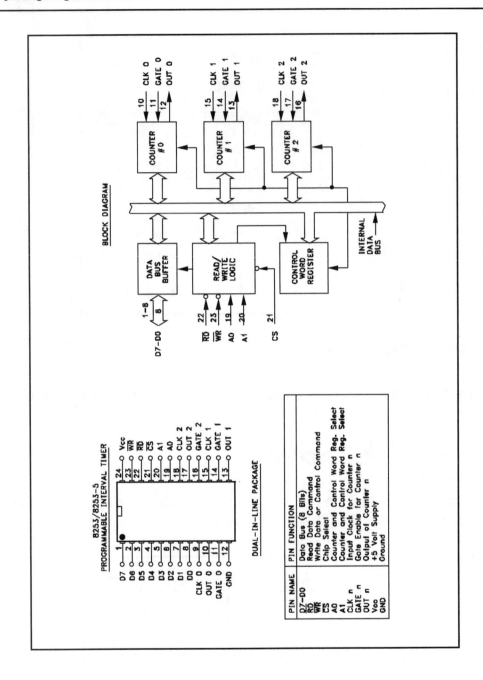

Figure 9.3. 8253/8253-5 package pinning, pin function definitions, and block diagram.

Data Bus Buffer

The device is provided with an eight-bit buffered bidirectional bus for data transfer. Data is transmitted or received by the buffer on the execution of INput or OUTput CPU instructions. The bus has three basic functions:

1. Programming the 8253 operating modes.
2. Loading the count registers.
3. Reading the count values.

Read/Write Logic

The read/write logic accepts inputs from the system bus. In turn, it generates control signals for overall device operation. It is enabled or disabled by the chip select (CS) input. Thus, changes are only possible when the device has been selected by the system logic.

Read (/RD)

(Note: In this book, "/" preceding a symbol indicates the complement.) This active low input informs the 8253 that the CPU is inputting (reading) data in the form of a counter's value.

Figure 9.4. 8253/8253-5 function selection logic. "No operation" refers to Read/Write of the data bus. Once a counter mode has been initiated, internal operations are maintained regardless of /CS status.

CS	RD	WR	A1	A0	Operation
0	1	0	0	0	Load Counter No. 0
0	1	0	0	1	Load Counter No. 1
0	1	0	1	0	Load Counter No. 2
0	1	0	1	1	Write MODE Word
0	0	1	0	0	Read Counter No. 0
0	0	1	0	1	Read Counter No. 1
0	0	1	1	0	Read Counter No. 2
0	0	1	1	1	No operation, 3-state
1	X	X	X	X	Disable 3-state
0	1	1	X	X	No operation, 3-state

Write (/WR)

This active low input informs the 8253 that the CPU is outputting (writing) data in the form of mode information or counter loading.

Addressing (A0, A1)

These inputs are normally connected to the address bus (Figure 9.1). Their function is the selection of one of the three counters for operation and to address the control word register for operating mode selection.

Chip Select (/CS)

A logic low on this input enables the 8253. No read or write operation is possible until the device has been selected. This input has no effect on the internal operation of the counters.

Figure 9.4 relates the chip select to the read, write, and address enabling logic.

The Control Word Register

The location of this register is seen in Figure 9.3. The register is selected when the address inputs A0,A1 are 11. It will then accept information from the data-bus buffer for register storage. This information controls the mode of each counter, selection of binary or BCD counting, and the count register loading. This register can only be written to; there is no provision for a status read of its content.

Counter #0, Counter #1, and Counter #2

The three functional blocks seen in Figure 9.3 are identical, so only the operation of a single counter is described.

Each counter consists of a single, 16-bit, presettable down counter. The counter can operate in binary or BCD. Its input, gate, and output are configured by the selection of modes stored in the control word register.

Each counter is fully independent and can have a separate mode configuration and counting operation. Special features are included for handling of the count values for minimizing the software requirements of these functions. The counter content is available to the CPU with simple read operations for event-counting applications. Special commands and logic included in the counter provide for reading values "on the fly" without having to inhibit the operation.

8253 Operational Description

General

The complete functional definition of the 8253 is programmed by the system's software. A set of control words *must* be sent out by the CPU to

Figure 9.5. 8253/8253-5 Control Word format with the function of each Bit defined. The Control Word must be entered on the Data Bus as a first step in initializing a counter.

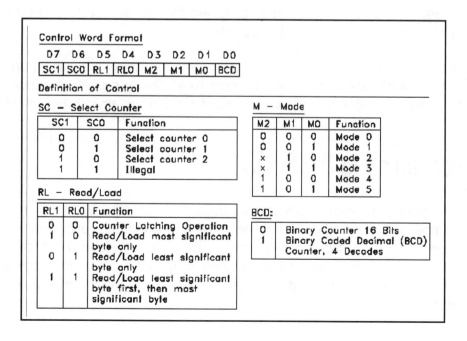

Modes	Signal Status	Low or Going Low	Rising	High
0		Disables Counting	---	Enables Counting
1		---	1) Initiates Counting 2) Resets after Next Clock	---
2		1) Disables Counting 2) Sets Output Immediately High	1) Reloads Counter 2) Initiates Counting	Enables Counting
3		1) Disables Counting 2) Sets Output Immediately High	1) Reloads Counter 2) Initiates Counting	Enables Counting
4		Disables Counting	---	
5		---	Initiates Counting	---

Figure 9.6. 8253/8253-5 Gate functions for each of the six modes of counter operation. Note that both level and rise/fall transition requirements exist.

initialize each counter with the desired mode and quantity information. Prior to device initialization, the mode, count, and output of all the counters is undefined. The control words program the mode, loading sequence, and selection of binary or BCD counting.

The actual counting operation of each counter is completely independent, and additional logic is provided on-chip to eliminate problems associated with efficient monitoring and management of external, asynchronous events or rates.

Programming the 8253

All the modes for each counter are programmed by the systems software by simple I/O operations. Each counter is individually programmed by writing into the control word register (A0,A1=11). *Figure 9.5* shows the format of the control word register. This figure also contains logic definition tables for programming the eight bits of the control word. These definitions follow.

SC - Select Counter

Counter selection is bits 7 and 6: SC1 and SC0, respectively, as seen in SC table of Figure 9.5.

RL - Read/Load

The read/load function logic is defined in the RL table of Figure 9.5. The count register is not loaded until the count value is written (one or two bytes, depending on the mode selected), followed by a rising edge and a falling edge of the clock. Any read of the counter prior to that falling edge might yield invalid data.

Mode Definition

Figure 9.6 is a summary of the gate pin operations for each of the six mode descriptions that follow. The gate input is meant by reference to a trigger in the text. Timing diagrams for each of the six modes are shown in *Figure 9.7*.

Mode 0: Interrupt on Terminal Count

The counter output will remain low following the mode set operation. When the count has been loaded into the selected count register, the output continues to remain low until the count has decremented to zero, at which time the output goes high.

The output will remain high until the selected count register is reloaded with the mode, or until a new count is loaded. The counter continues to decrement after terminal count has been reached.

Rewriting a counter register during counting results in the following.

1. Write of the first byte stops the current counting.
2. Write of the second byte starts the new count.

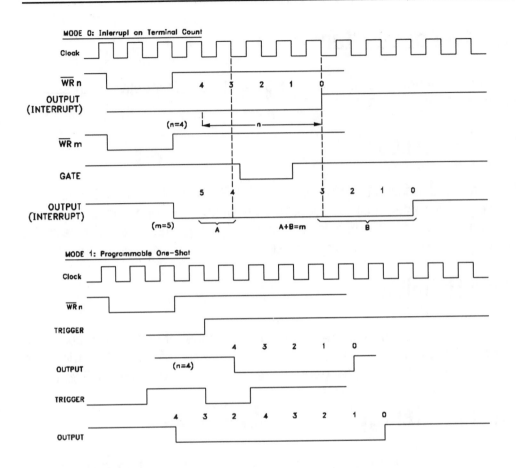

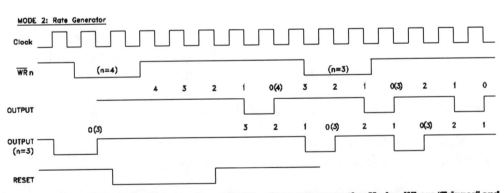

Figure 9.7. 8253/8253-5 timing diagrams for each of the six counter operation Modes. Where "Trigger" and "Reset" are given, the "Gate" is meant. The Gate is held at logic level 1 for Mode 3. *(Continued on next page.)*

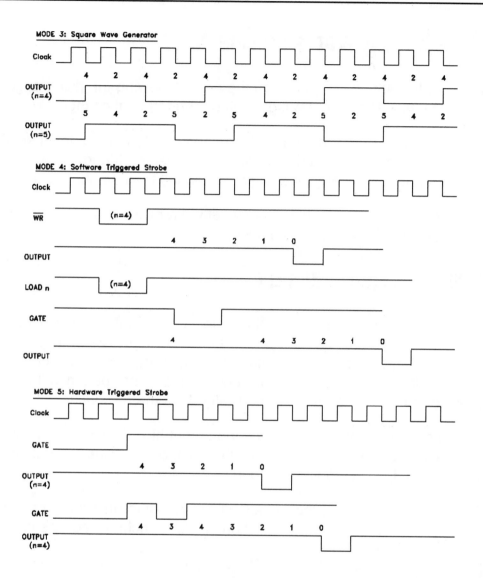

MODE 3: Square Wave Generator

MODE 4: Software Triggered Strobe

MODE 5: Hardware Triggered Strobe

Mode 1: Programmable One-Shot

The counter output will go low on the count following the rising edge of the gate input. The output will go high on the terminal count.

If a new count value is loaded while the output is low, it will not affect the duration of the one-shot pulse until the succeeding trigger. The current count can be read at any time without affecting the one-shot pulse.

The one-shot is retriggerable. The output will remain low for the full count after any rising edge of the gate input.

Mode 2: Rate Generator

This mode is that of a divide-by-N counter. The output will be low for one period of the input clock. The period from one output pulse to the next equals the number of input counts in the count register. If the count register is reloaded between output pulses the present period will not be affected, but the subsequent period will reflect the new value.

The gate input, when low, will force the output high. When the gate input goes high, the counter will start from the initial count. Thus, the gate input can be used to synchronize the counter.

When this mode is set, the output will remain high until after the count register is loaded. The output can then also be synchronized by software.

Mode 3: Square-Wave Generator

This mode is similar to Mode 2, except the output will remain high until one-half the count has been completed (for even numbers) and go low for the other half of the count.

This is accomplished by de-crementing the counter by two on the falling edge of each clock pulse. When the counter reaches terminal count, the state of the output is changed and the counter is reloaded with the full count and the whole process repeated.

If the count is odd and the output is high, the first clock pulse (after the count is loaded) decrements the count by 1. Subsequent clock pulses decrement the count by 2. After timeout, the output pulse goes low and the full count is reloaded. The first clock pulse (following the reload) decrements the counter by 3.

Subsequent clock pulses decrement the counter by 2 until timeout. Then the whole process is repeated. In this way, if the count is odd, the count will be high for (N+1)/2 counts and low for (N-1)/2 counts. In modes 2 and 3, if a clock source other than the system clock is used, gate should be pulsed immediately following the write (/WR) of a new count value.

Mode 4: Software-Triggered Strobe

After the mode is set, the output will go high. When the count is loaded, the counter will begin counting. On the terminal count, the output will go low for one input clock period, then will go high again. If the count register is reloaded during counting, the new count will be loaded on the next CLK pulse. The count will be inhibited while the gate input is low.

Mode 5: Hardware-Triggered Strobe

The counter will start counting after the rising edge of the trigger input and will go low for one clock period when the terminal count is reached. The counter is retriggerable. The output will not go low until the full count after the rising edge of any trigger.

The 8253 Read/Write Procedure

Write Operations

The systems software must program each counter with the mode and quantity desired. The programmer must write out a mode control word and the programmed number of count register bytes (one or two) prior to actually using the selected counter.

The actual order of the programming is made flexible. Writing out of the mode control word can be in any sequence of counter selection (e.g., Counter 0 does not have to be first or Counter 2 last.) Each counter's mode control word register has a separate address that makes its loading completely sequence independent (SC0,SC1).

The loading of the count register with the actual count value, however, must be done in exactly the same sequence programmed in the mode control word (RL0,RL1). This loading of the counter's count register is still sequence independent like the mode control word loading. But when a selected count register is to be loaded, it *must* be loaded with the number of bytes programmed in the mode control word (RL0,RL1). The one or two bytes to be loaded in the count register do not have to follow

Figure 9.8(a) & (b). 8253/8253-5 Control Word and Mode programming format. No requirement for a specific programming sequence exists. One, two, or all three counters may be programmed for operation as required.

		A1	A0
No. 1	MODE Control Word Counter 0	1	1
No. 2	MODE Control Word Counter 1	1	1
No. 3	MODE Control Word Counter 2	1	1
No. 4	LSB Count Register Byte Counter 1	0	1
No. 5	MSB Count Register Byte Counter 1	0	1
No. 6	LSB Count Register Byte Counter 2	1	0
No. 7	MSB Count Register Byte Counter 2	1	0
No. 8	LSB Count Register Byte Counter 0	0	0
No. 9	MSB Count Register Byte Counter 0	0	0

Programming format table:

MODE Control Word Counter n
LSB Count Register Byte Counter n
MSB Count Register Byte Counter n

(a) Programming Format

(b) Alternate Programming Formats

the associated mode control word; they can be programmed at any time following the mode control word loading, as long as the correct number of bytes is loaded in order.

All counters are down counters. Thus, the value loaded into the count register will actually be decremented. Loading all zeroes into a count register will result in the maximum count ($2e16$ for Binary, $10e4$ for BCD). In Mode 0, the new count will not restart until the load has been completed. It will accept one or two bytes—depending on how the mode control words (RL0,RL1) are programmed—then proceed with the restart operation.

Figures 9.8(a) and *9.8(b)* define two programming formats. The format in (a) is a simple example of loading the 8253 and does not imply that it's the only format used. In (b), the exclusive addresses of each counter's count register make the task of programming the 8253 a very simple matter, and maximum effective use of the device will result if this feature is fully utilized.

Read Operations

In counter operations, it often becomes necessary to read the value of the count in progress and make a computational decision based on its quantity. Event counters are a common application that use this function. The 8253 uses logic that will allow the programmer to readily read the content of any of the three counters without disturbing the actual count in progress.

Figure 9.9. 8253/8253-5 Read operation addressing. The Read and Write addressing is identical. The Read input, /RD, must be Low. No read of the Control Register is possible.

A1	A0	RD	
0	0	0	Read Counter No. 0
0	1	0	Read Counter No. 1
1	0	0	Read Counter No. 2
1	1	0	Illegal

Read Operation Chart

```
AO, A1 = 11

┌───┬───┬───┬───┬───┬───┬───┬───┐
│ D7│ D6│ D5│ D4│ D3│ D2│ D1│ D0│
├───┼───┼───┼───┼───┼───┼───┼───┤
│SC1│SC0│ 0 │ 0 │ x │ x │ x │ x │
└───┴───┴───┴───┴───┴───┴───┴───┘

SC1,SC0  — specify counter to be latched.
D5,D4    — 00 designates counter latching
           operation.
x        — Don't care
```

Figure 9.10. 8253/8253-5 Mode Register format for latching a counter count. The command has no effect on the counter's operating mode.

There are two methods the programmer can use to read the value of the counters. The first involves the use of simple I/O read operations of the selected counter. By controlling the A0,A1 inputs to the device, the programmer can select the counter to be read (remember that no read operation of the mode register is allowed). The only requirement with this method is that, in order to assure a stable count reading, the actual operation of the selected counter *must be inhibited* either by controlling the gate input or by external logic that inhibits the clock input. The content of the counter selected will be as follows.

• First I/O read contains the least significant bit (LSB).

• Second I/O read contains the most significant bit (MSB).

Due to the internal logic of the 8253, it is absolutely necessary to complete the entire reading procedure. If two bytes are programmed to be read, then two bytes *must* be read before any loading (/WR) command can be sent to the same counter.

Reading While Counting

In order for the programmer to read the content of any counter without affecting or disturbing the counting operation, the 8253 has special internal logic that can be accessed using simple WR commands to the mode register.

Bus Parameters *

READ CYCLE

Symbol	Parameter	8253		8253-5		Unit
		Min.	Max.	Min.	Max.	
tAR	Address Stable Before READ	50		30		nS
tRA	Address Hold Time For READ	5		5		nS
tRR	READ Pulse Width	400		300		nS
tRD	Data Delay From READ **		300		200	nS
tDF	READ to Data Floating	25	125	25	100	nS
tRV	Recovery Time Between READ and Any Other Control Signal	1		1		uS

WRITE CYCLE

Symbol	Parameter	8253		8253-5		Unit
		Min.	Max.	Min.	Max.	
tAW	Address Stable Before WRITE	50		30		nS
tWA	Address Hold Time for WRITE	30		30		nS
tWW	WRITE Pulse Width	400		300		nS
tDW	Data Set Up Time for WRITE	300		250		nS
tWD	Data Hold Time for WRITE	40		30		nS
tRV	Recovery Time Between WRITE and Any Other Control Signal	1		1		uS

CLOCK AND GATE TIMING

Symbol	Parameter	8253		8253-5		Unit
		Min.	Max.	Min.	Max.	
tCLK	Clock Period	380		380		nS
tPHW	High Pulse Width	230		230		nS
tPWL	Low Pulse Width	150		150		nS
tGW	Gate Width High	150		150		nS
tGL	Gate Width Low	100		100		nS
tGS	Gate Set Up Time to Clock↑	100		100		nS
tGH	Gate Hold Time After Clock↑	50		50		nS
tOD	Output Delay From Clock **		400		400	nS
tODG	Output Delay From Gate **		300		300	nS

NOTES; * AC timings measured at VOH=2.2, VOL=0.8
       **CL = 150pf

Table 9.1. 8253/8253-5 AC characteristics.

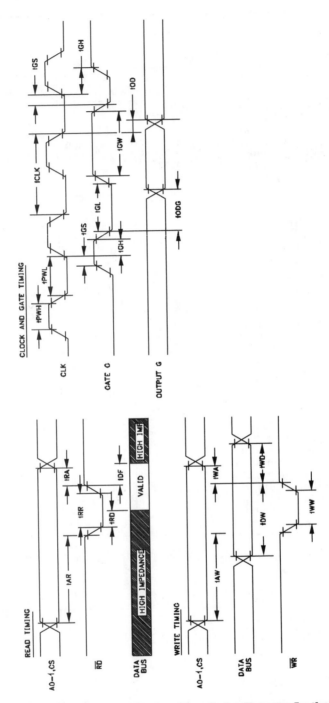

Figure 9.11. 8253/8253-5 Read, Write, and Clock and Gate timing diagrams. For the bench operation, the Read input is a level to provide time for viewing.

Basically, when the programmer wishes to read the content of a selected register "on the fly," he loads the mode register with a special code that latches the count into a storage register so that its value contains an accurate stable quantity. The programmer then issues a read command to the selected counter, and the content of latched register is available. *Figure 9.9* is a read operations chart for the counter to be selected.

Mode Register for Latching Count

Figure 9.10 describes the selection requirements of the counter whose data is to be latched. This function is the counter latch command. It, too, is written to the control word register, selected by A0,A1 = 11. The command is identified by the two bits (D5,D4) set to 0,0.

The counter whose data is to be latched is selected by SC1,SC2 as with the control word. The selected counter's output latch (OL) latches the count at the time the latch command is received. The count is held until the latch is read by the CPU, or until the latch is reprogrammed. The latch is then automatically unlatched and the OL returns to following the counter.

The same limitation applies to this mode of reading the counter value as previously described. It is mandatory to complete the entire operation as programmed. This command has no effect on the counter's mode.

8253 Timing and AC Characteristics

Table 9.1 defines the AC characteristics for the 8253 and the 8253-5. *Figure 9.11* provides timing diagrams for read, write, and clock and gate timing.

Bench-Top Operation of the 8253/8253-5 Programmable Interval Timer

The 8253/8253-5 programmable interval timer is a departure from the data-communication devices of the previous chapters. As we can see from the

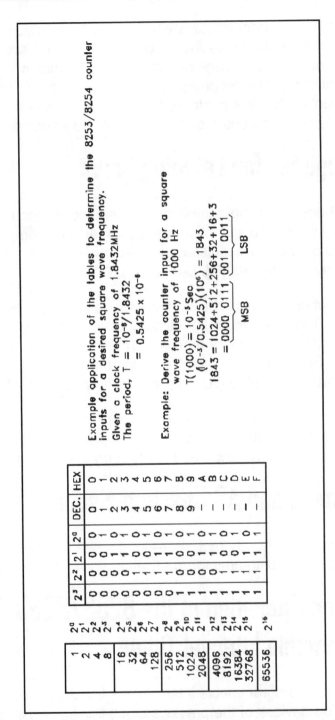

Example application of the tables to determine the 8253/8254 counter inputs for a desired square wave frequency.

Given a clock frequency of 1.8432MHz
The period, T = $10^{-6}/1.8432$
         = 0.5425×10^{-6}

Example: Derive the counter input for a square wave frequency of 1000 Hz

$T(1000) = 10^{-3}$ Sec
$(10^{-3}/0.5425)(10^6) = 1843$
$1843 = 1024+512+256+32+16+3$
      = 0000 0111 0011 0011

$\underbrace{\text{0000 0111}}_{\text{MSB}} \quad \underbrace{\text{0011 0011}}_{\text{LSB}}$

	2^3	2^2	2^1	2^0	DEC.	HEX
2^0	0	0	0	0	0	0
2^1	0	0	0	1	1	1
2^2	0	0	1	0	2	2
2^3	0	0	1	1	3	3
2^4	0	1	0	0	4	4
2^5	0	1	0	1	5	5
2^6	0	1	1	0	6	6
2^7	0	1	1	1	7	7
2^8	1	0	0	0	8	8
2^9	1	0	0	1	9	9
2^{10}	1	0	1	0	—	A
2^{11}	1	0	1	1	—	B
2^{12}	1	1	0	0	—	C
2^{13}	1	1	0	1	—	D
2^{14}	1	1	1	0	—	E
2^{15}	1	1	1	1	—	F

1	
2	
4	
8	
16	
32	
64	
128	
256	
512	
1024	
2048	
4096	
8192	
16384	
32768	
65536	2^{16}

Table 9.2. Binary and BCD tables, with a worked example for a counter input to provide a square-wave output.

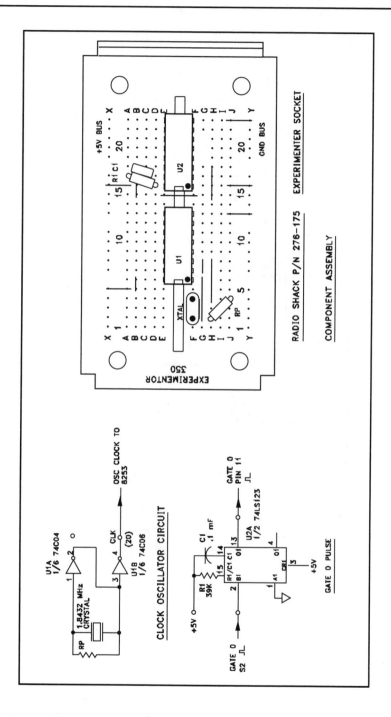

Figure 9.12. The 1.8432-MHz clock source and Gate 0 pulse source circuitry and prototype assembly used with the bench-top operation of the 8253/8253-5.

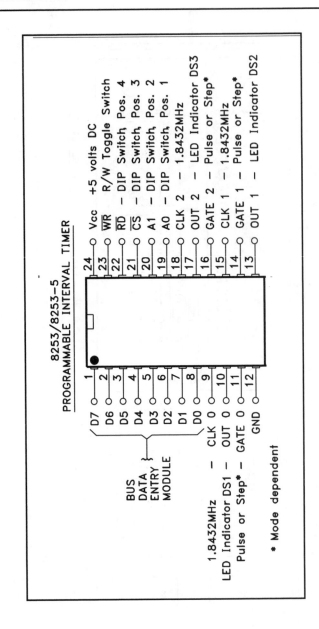

Figure 9.13. A helpful planning diagram for wiring the bench-top configuration of the 8253/8253-5.

bench-top assembly drawings, it is a simpler and easier device to configure for operation. Operation is also very straightforward, as we will discover from the procedures provided.

There is, however, a complexity of another sort in the programming of the counters. In this, we have the option of binary or BCD notation. *Table 9.2* provides a binary table, plus an example of a square-wave frequency application.

Let's consider the basics now. The clock source is a frequency of 1.8432 MHz. The period of one clock cycle is:

10(e-6)/1.8432 = 0.5425(10e-6) = 0.5425 μS

Now suppose we want to use counter 1 in the one-shot mode with a pulse width of 330 μS. The ratio of 330/.5425 = 608.3, rounded off to 608. Let's take advantage of the binary table:

608 = 512+64+32
     = 0000+0010+0110+0000
in binary notation, from which the MSB is 0000 0010 and the LSB is 0110 0000.

Next, we enter these values, using the one-shot procedure of Figure 9.16. Upon triggering Gate 1, we see a negative-going pulse of 330 μS, just as we would expect. But first we will have to learn how we get to this point in our experience.

Figure 9.12 describes the prototype clock circuitry and assembly employed in the previous chapters. So you may have made use of it already. There is a change in the monostable circuit in that only one is used, designated for the gate input of Counter 0. There is no change in the clock.

Figure 9.13 is a planning diagram for wiring the 8253 in the bench-top configuration. We will use it as a guide in the wiring of the operating setup, as shown in *Figure 9.14*.

Earlier chapters have provided suggestions for the construction of the modules and their corrugated cardboard supports, so these are not repeated

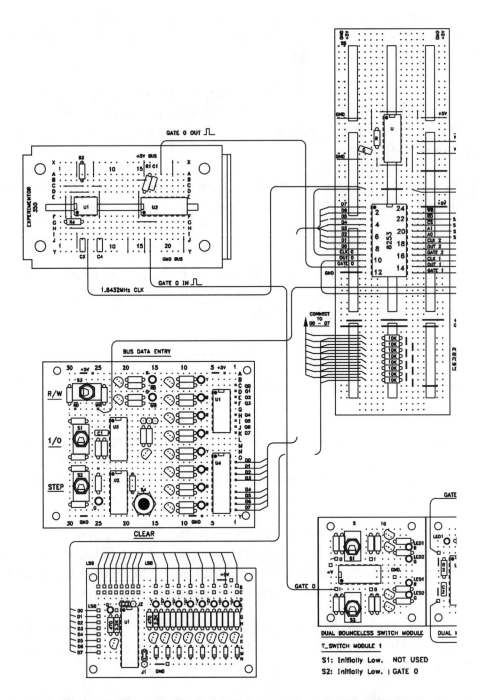

Figure 9.14. The author's assembly and wiring for bench-top operation of the 8253/8253-5. *(Continued on next page.)*

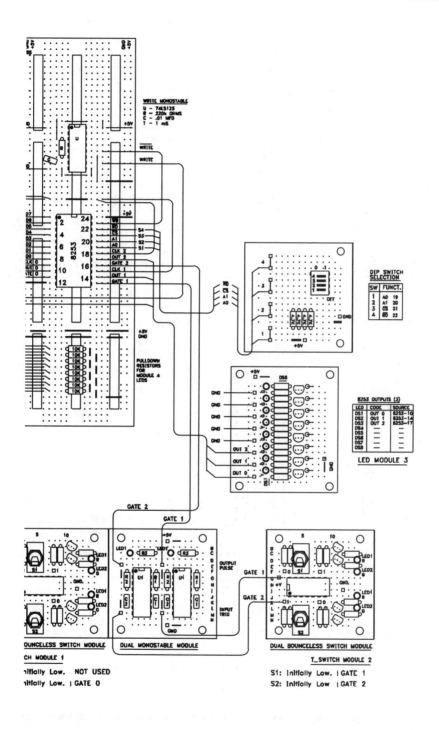

PROCEDURE FOR 8253 OPERATION AS A SQUARE WAVE GENERATOR

NOTES: The Bus Data Entry Toggle Switch is to remain in the READ
Position except when toggled for a WRITE operation.
DIP Switch #4 to remain in the 1 position except when a Data
Bus READ is desired.
For this MODE Gate 0 is to be tied to the +5V bus.
An oscilliscope and frequency meter are required to properly
observe the square wave timing and frequency.

1. Setup Counter 0 for the Square Wave, MODE 3. ☐

OFF

1 A0			1
2 A1			1
3 $\overline{CS}$			0
4 $\overline{RD}$			1 ☐

1 0

DIP SW SETTINGS

BUS DATA ENTRY = 0011 0110 ☐

7 6 5 4 3 2 1 0	OPERATION
0 0 1 1 0 1 1 0	Counter and MODE select

└── 16-Bit Binary Counter
Mode 3 select
Load LSB first, then MSB
Counter 0 select

Toggle Data Entry Switch to Write ☐

2. Load Counter 0 with the Least Significant Bits (LSB) ☐

OFF

1 A0			0
2 A1			0
3 $\overline{CS}$			0
4 $\overline{RD}$			1 ☐

1 0

DIP SW SETTINGS

BUS DATA ENTRY = 1010 1110 ☐

7 6 5 4 3 2 1 0	OPERATION
1 0 1 0 1 1 1 0	Least Significant Byte (LSB)

└── Binary 8+4+2
Binary 128+32

Note: Binary values shown are suggested only.
Toggle Data Entry Switch to Write ☐

3. Load Counter 0 with the Most Significant Bits (MSB) ☐

OFF

1 A0			0
2 A1			0
3 $\overline{CS}$			0
4 $\overline{RD}$			1 ☐

1 0

DIP SW SETTINGS

BUS DATA ENTRY = 0000 1101 ☐

7 6 5 4 3 2 1 0	OPERATION
0 0 0 0 1 1 0 1	Most Significant Byte (MSB)

└── Binary 2048+1024+256
Binary 0+0+0+0

Note: Binary values shown are suggested only.
The frequency is 520 Hz for the counter values
given.
Toggle Data Entry Switch to Write ☐

4. Load the counters, LSB and MSB, with all ones to observe a flickering
of LED DS1, displaying OUT 0. With these values f~ 29Hz, T ~ 34mS.
Repeat steps 2 and 3 with new values. Keep in mind the LSB and MSB
must both be entered, even if the MSB is all zeroes.

Figure 9.15. The procedure for bench-top operation of the 8253/8253-5 as a square-wave generator using counter 0.

PROCEDURE FOR 8253 OPERATION AS A ONE-SHOT

NOTE: The Bus Data Entry Toggle Switch is to remain in the READ
Position except when toggled for a WRITE Operation.
DIP Switch #4 to remain in the 1 position except when a Data
Bus READ is desired.

1. Setup Counter 1 for the One-Shot, MODE 1. ☐

BUS DATA ENTRY = 0111 0010 ☐

7 6 5 4 3 2 1 0	OPERATION
0 1 1 1 0 0 1 0	Counter and MODE select

OFF

1 A0	1
2 A1	1
3 C̄S̄	0
4 R̄D̄	1

1 0

DIP SW SETTINGS

— 16-Bit Binary Counter
— Mode 1 Select
— Load LSB first, then MSB
— Counter 1 Select

Toggle Data Entry Switch to Write

2. Load Counter 1 with the Least Significant Byte (LSB) ☐

BUS DATA ENTRY = 1110 0100 ☐

7 6 5 4 3 2 1 0	OPERATION
1 1 1 0 0 1 0 0	Least Significant Byte (LSB)

OFF

1 A0	1
2 A1	0
3 C̄S̄	0
4 R̄D̄	1

1 0

DIP SW SETTINGS

— Binary 4
— Binary 128+64+32

Note: Binary values shown are suggested only.
Toggle Data Entry Switch to Write

3. Load Counter 1 with the Most Significant Byte (MSB) ☐

BUS DATA ENTRY = ☐

7 6 5 4 3 2 1 0	OPERATION
0 0 0 0 1 1 0 1	Most Significant Byte (MSB)

OFF

1 A0	1
2 A1	0
3 C̄S̄	0
4 R̄D̄	1

1 0

DIP SW SETTINGS

— Binary 2048+1024+256
— Binary 0+0+0+0

Note: Binary values shown are suggested only.
The pulse width is 2mS for the counter values given.
Toggle Data Entry Switch to Write

4. Toggle switch for GATE 1 to trigger the one-shot while observing OUT 1
on the oscilloscope and/or LED DS 2.
Note that DS2 is normally ON. It will blink when a sufficiently long
pulse is triggered.
Repeat steps 2 and 3 with new values. Keep in mid the LSB and MSB
must both be entered even if the MSB is all zeroes.

Figure 9.16. The procedure for bench-top operation of the 8253/8253-5 as a one-shot using counter 1.

PROCEDURE FOR 8253 OPERATION AS A RATE GENERATOR

NOTES: The Bus Data Entry Toggle Switch is to remain in the READ
Position except when toggled for a WRITE operation.
DIP Switch #4 to remain in the 1 position except when a Data
Bus READ is desired.
For this mode use T-Switch Module 2, S2 for DC Gate Control.
The GATE Input must be High to observe the rate pulse at the
counter 2 output. The switch may be toggled to suspend the output.
An oscilloscope and frequency meter are required to properly
observe the pulse period and frequency.

1. Setup Counter 2 for the Rate Generator, MODE 2. ☐

BUS DATA ENTRY = 1001 0100 ☐

OFF

1	A0	1
2	A1	1
3	CS̄	0
4	RD̄	1

1   0
DIP SW SETTINGS

7 6 5 4 3 2 1 0	OPERATION
1 0 0 1 0 1 0 0	Counter and MODE select

— 16-Bit Binary Counter
Mode 2 select
Read/Load LSB only
Counter 2 select

Toggle Data Entry Switch to Write ☐

2. Load Counter 2 with the Least Significant Byte (LSB) ☐

BUS DATA ENTRY = 1100 1011 ☐

OFF

1	A0	0
2	A1	1
3	CS̄	0
4	RD̄	1

1   0
DIP SW SETTINGS

7 6 5 4 3 2 1 0	OPERATION
1 1 0 0 1 0 1 1	Least Significant Byte (LSB)

— Binary 8+3
— Binary 128+64

Note: Binary values shown are suggested only.
Toggle Data Entry Switch to Write ☐
The rate period is 110 uS for the values entered.

3. Load Counter 2 with the Most Significant Byte (MSB) ☐

BUS DATA ENTRY = ☐

OFF

1	A0	0
2	A1	1
3	CS̄	0
4	RD̄	1

1   0
DIP SW SETTINGS

7 6 5 4 3 2 1 0	OPERATION
	Most Significant Byte (MSB)

Note: This step left blank for your entry if desired.
Revise Control Word for LSB and MSB.
Toggle Data Entry Switch to Write ☐

4. Repeat step 2 with new values. Vary the pulse timing using
Decimal value = Period $(10^{-6})/0.5425\ (10^{-6})$.

Figure 9.17. The procedure for bench-top operation of the 8253/8253-5 as a rate generator using counter 2.

here. Do take care with the circuit connections, double-checking as you go, as with the congested layout it is easy to poke a wire into an incorrect point.

As with devices from the previous chapters, we are fortunate in that we are able to input data on the bus at a slow rate even though operation is with a high-speed clock. With the 8253, it's convenient to employ a monostable to clock in the data. We cannot take a similar route with reading the bus, however, as the data is not internally latched in these procedures. A DIP switch is used for /RD to enable reading a counter's status. Remember, though, that the control word register cannot be read.

Three test procedures are provided here, one for each counter. They are described in *Figures 9.15, 9.16,* and *9.17.* The procedure of Figure 9.15 uses counter 0 as a square-wave generator. For this function, we must connect the gate input to the +5V bus. The monostable connection shown in the wiring is in the event that you wish to experiment with other applications of this counter.

Let's see where the binary values for the square-wave came from.

1/520 = 1900 mS
1.900(10e-3)/.5425(10e-6) =3502
From the binary table,
$$3502=2048+1024+256+128+32+8+4+2$$
$$=0000+1101+0011+0110$$
which is as shown in the procedure.

The procedure of Figure 9.16 uses counter 1 as a one-shot. In this, we have worked out the math for a one-shot timing that differs from the example. The procedure of Figure 9.17 uses counter 2 as a rate generator. The counter output is high with the gate input low. If the gate is held high continuously, a negative-going pulse will be outputted with a pulse width of the clock period. If the gate input is lowered, the pulses will cease, but raising the gate will restore them. Thus, the gate can be used as a synchronizer.

The procedures are designed to be self-explanatory as much as possible, but it will pay to review the text before proceeding with their operation. You may have difficulties, but perseverance pays dividends in satisfaction. After gaining experience with the examples given, you should experiment with other values for these modes and then venture forth with some of the other modes as well.

References

1. Intel Corporation, *Microsystem Components Handbook, Vol. 2,* Data Sheet, "8253/8253-5 Programmable Interval Timer," 1984, p. 6-331.

2. Ibid, Data Sheet, "8205 High Speed 1 Out of 8 Binary Decoder," 1984, p. 2-50. (*MCS* is a registered trademark of Intel Corporation.)

3. Reference for Figure 9.1: Intel Corporation, *Microsystem Components Handbook, Vol. 2,* Data Sheet, "8253/8253-5 Programmable Interval Timer," 1984, p. 6-333.

4. Reference for Figure 9.2: Ibid. Figure 10, p. 6-338.

5. Reference for Figure 9.3: Ibid. Figures 1 and 2, p. 6-331.

6. Reference for Figure 9.4: Ibid. Address Table, p. 6-332.

7. Reference for Figure 9.5: Ibid. Control Definition Tables, p. 6-334.

8. Reference for Figure 9.6:Ibid. Figure 6, p. 6-335.

9. Reference for Figure 9.7: Ibid. Figure 7, p. 6-336.

10. Reference for Figure 9.8: Ibid. Figures 8 and 9, p. 6-337.

11. Reference for Figure 9.9: Ibid. Read Operation Chart, p. 6-338.

12. Reference for Figure 9.10: Ibid. Latching Count Register, p. 6-338.

13. Reference for Figure 9.11: Ibid. Waveforms, p. 6-341.

14. Reference for Table 9.1: Ibid. AC Characteristics, p. 6-339, 6-340.

Chapter 10
The 8254/8254-2
Programmable Interval Timer

Introduction

The 8254[1] is a programmable interval timer/counter specifically designed for use with the Intel microcomputer system. Its function is that of a general purpose, multitiming element that can be treated as an array of I/O ports in the system software. In many respects, operation of the 8254 duplicates that of the 8253 with additional features.

In its system application, the 8254 generates accurate time delays under software control. The programmer configures the 8254 to match the requirements, initializing one of the counters with the required quantity. Upon command, the device will count out the delay and interrupt the CPU when its purpose has been accomplished. Multiple delays can be realized by an assignment of priorities.

Other counter/timer functions that are nondelay common to most microcomputers can be implemented with the 8254. These include:

- Programmable rate generator
- Event counter
- Binary rate multiplier
- Real-time clock
- Digital one-shot
- Complex motor controller
- Complex waveform generator

The 8254 System Interface

Figure 10.1 illustrates the device interface in a microcomputer system. The 8254 is a component of the Intel microcomputer system and interfaces in the same manner as other components of the system.

The system software treats it as an array of peripheral I/O ports; three are counters and a fourth is a control register for mode programming. The select inputs A0,A1 connect to the A0,A1 address bus signals of the CPU. The /CS can be derived directly from the address bus using a linear select

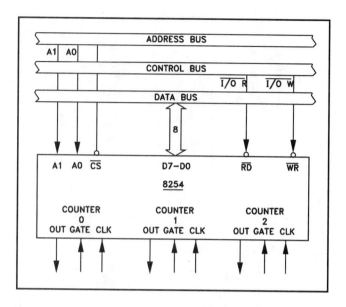

Figure 10.1. The 8254 System Interface.

method, or, alternatively, to the output of a decoder, such as the Intel 8205 for larger systems.

Functional Description

Figure 10.2 describes the package pinning, the pin functions, and provides a block diagram of the internal configuration.

Data Bus Buffer

The device is provided with an eight-bit buffered bidirectional bus for data transfer. Data is transmitted or received by the buffer on the execution of INput or OUTput CPU instructions. The bus has three basic functions:

1. Programming the 8254 operating modes.

2. Loading the count registers.

3. Reading the count values.

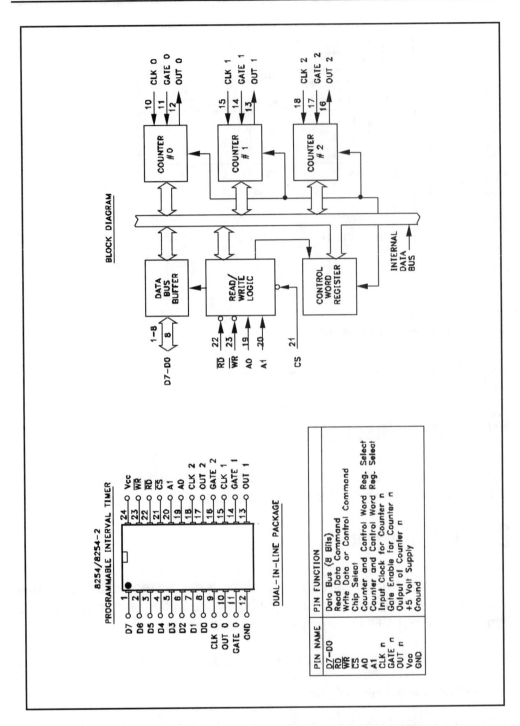

Figure 10.2. 8254/8254-2 package pinning, pin function definitions, and block diagram.

Read/Write Logic

The read/write logic accepts inputs from the system bus. In turn, it generates control signals for the other functional blocks of the 8254. It is enabled or disabled by the chip select (CS) input. Thus, changes are only possible when the device has been selected by the system logic.

Read (/RD)

(Note: In this book, "/" preceding a symbol indicates the complement.) This active low input informs the 8254 that the CPU is inputting (reading) data in the form of a counter's value.

Write (/WR)

This active low input informs the 8254 that the CPU is outputting (writing) data in the form of mode information or counter loading.

Addressing (A0,A1)

These inputs are normally connected to the address bus (Figure 10.1). Their function is the selection of one of the three counters for operation and to address the control word register for operating mode selection.

$\overline{CS}$	$\overline{RD}$	$\overline{WR}$	A1	A0	Operation
0	1	0	0	0	Write into Counter 0
0	1	0	0	1	Write into Counter 1
0	1	0	1	0	Write into Counter 2
0	1	0	1	1	Write Control Word
0	0	1	0	0	Read from Counter 0
0	0	1	0	1	Read from Counter 1
0	0	1	1	0	Read from Counter 2
0	0	1	1	1	No-Operation (3-State)
1	X	X	X	X	No-Operation (3-State)
0	1	1	X	X	No-Operation (3-State)

Figure 10.3. 8254/8254-2 function selection logic. "No operation" refers to read/write of the data bus. Once a counter mode has been initiated, internal operations are maintained regardless of /CS status.

Chip Select (/CS)

A logic low on this input enables the 8254. No read or write operation is possible until the device has been selected. This input has no effect on the internal operation of the counters. *Figure 10.3* relates the chip select to the read, write, and address enabling logic.

The Control Word Register

The location of this register is seen in Figure 10.2. The register is selected when the address inputs A0,A1 are 11. It will then accept information from the data-bus buffer for register storage. This information controls the mode of each counter, selection of binary or BCD counting, and the count register loading.

The register is not a part of the counter itself, but its content determines how the counter operates. This register can only be written to; status information is available, however, with the read-back command.

Counter #0, Counter #1, and Counter #2

The three functional blocks seen in Figure 10.2 are identical, so only the operation of a single counter is described. *Figure 10.4* illustrates the internal functioning of the counters. This figure shows us the relationship between the control logic, the control word and status registers, the status latch, and the counting element (CE).

Referring to the figure, we see that the status register, when latched, contains the current contents of the control word register and status of the output and null count flag. (This is explained in detail later when describing the read-back command.)

CE is the actual counter consisting of a single, 16-bit, presettable synchronous down counter. The counter can operate in binary or BCD. Its

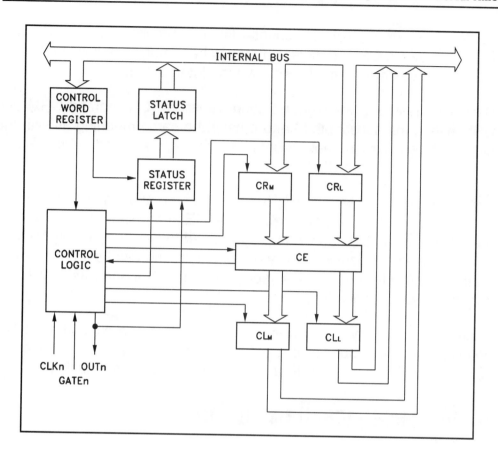

Figure 10.4. Internal block diagram of the 8254 counter.

input, gate, and output are configured by the selection of modes stored in the control word register. Note that the CE itself cannot be read; it is the output latches (OL) that are being read. Nor is it written into; it is the count registers (CR) that are written to.

OLM and OLL are two eight-bit latches where "M" and "L" stand for "most" and "least" significant bytes, respectively. These are spoken of as a single unit, the OL. The latches follow the CE in normal operation, but latch the count data upon receipt of a counter-latch command. The count is held until read by the CPU, after which the latches return to following the CE. One latch at a time is enabled to drive the internal bus.

Similarly, there are two eight-bit count registers (CR) named CRM and CRL, where the subscripts M and L have the same meaning as with the output latches. When a new count is entered in, it's stored in the CR and later transferred to the CE.

The control logic allows one register to be loaded at a time from the internal bus. Both bytes are transferred to the CE at the same time. The CRs are cleared when the counter is programmed to assure that only the current status—least significant only or both least and most significant—has been entered. If the counter has been programmed for one byte only (LSB), the other bytes will be all zeroes.

Each counter is fully independent, with its own mode configuration and counting operation. Special features are included for handling of the count values for minimizing the software requirements of these functions.

8254 Operational Description

General

The complete functional definition of the 8254 is programmed by the system's software. A set of control words *must* be sent out by the CPU to initialize each counter with the desired mode and quantity information. Prior to device initialization, the mode, count value, and output of all the counters is undefined. The control words program the mode, loading sequence, and selection of binary or BCD counting. Unused counters need not be programmed.

The actual counting operation of each counter is completely independent, and additional logic is provided on-chip to eliminate problems associated with efficient monitoring and management of external, asynchronous events or rates.

```
Control Word Format

D7   D6   D5   D4   D3   D2   D1   D0
┌────┬────┬────┬────┬────┬────┬────┬────┐
│SC1 │SC0 │RL1 │RL0 │ M2 │ M1 │ M0 │BCD │
└────┴────┴────┴────┴────┴────┴────┴────┘
Definition of Control
```

SC — Select Counter

SC1	SC0	Function
0	0	Select counter 0
0	1	Select counter 1
1	0	Select counter 2
1	1	Read-Back Command

M — Mode

M2	M1	M0	Function
0	0	0	Mode 0
0	0	1	Mode 1
x	1	0	Mode 2
x	1	1	Mode 3
1	0	0	Mode 4
1	0	1	Mode 5

RW — Read/Write

RL1	RL0	Function
0	0	Counter Latching Operation
1	0	Read/Write most significant byte only
0	1	Read/Write least significant byte only
1	1	Read/Write least significant byte first, then most significant byte

BCD:

0	Binary Counter 16 Bits
1	Binary Coded Decimal (BCD) Counter, 4 Decades

Figure 10.5. 8254/8254-2 Control Word format with the function of each Bit defined. The Control Word must be entered on the Data Bus as a first step in initializing a counter.

Programming the 8254

All the modes for each counter are programmed by the systems software through simple I/O operations.

Each counter is individually programmed by writing into the control word register. The control word format is A1,A0=11 /CS=0 /RD=1 /WR=0. The programming operations follow.

SC - Select Counter

Counter selection is bits 7 and 6—SC1 and SC0, respectively—as seen in the SC table of *Figure 10.5*.

		A1	A0
Control Word — Counter 0		1	1
LSB of Count — Counter 0		0	0
MSB of Count — Counter 0		0	0
Control Word — Counter 1		1	1
LSB of Count — Counter 1		0	1
MSB of Count — Counter 1		0	1
Control Word — Counter 2		1	1
LSB of Count — Counter 2		1	0
MSB of Count — Counter 2		1	0

		A1	A0
Control Word — Counter 2		1	1
Control Word — Counter 1		1	1
Control Word — Counter 0		1	1
LSB of Count — Counter 2		1	0
MSB of Count — Counter 2		1	0
LSB of Count — Counter 1		0	1
MSB of Count — Counter 1		0	1
LSB of Count — Counter 0		0	0
MSB of Count — Counter 0		0	0

		A1	A0
Control Word — Counter 0		1	1
Control Word — Counter 1		1	1
Control Word — Counter 2		1	1
LSB of Count — Counter 2		1	0
LSB of Count — Counter 1		0	1
LSB of Count — Counter 0		0	0
MSB of Count — Counter 0		0	0
MSB of Count — Counter 1		0	1
MSB of Count — Counter 2		1	0

		A1	A0
Control Word — Counter 1		1	1
Control Word — Counter 0		1	1
LSB of Count — Counter 1		0	1
Control Word — Counter 2		1	1
LSB of Count — Counter 0		0	0
MSB of Count — Counter 1		0	1
LSB of Count — Counter 2		1	0
MSB of Count — Counter 0		0	0
MSB of Count — Counter 2		1	0

Figure 10.6. A few programming sequence examples. Note that in these all counters are programmed to read/write two-byte counts. These are but four of many possible one- or two-byte programming sequences.

RW - Read/Write

The read/write function logic is defined in the RW table of Figure 10.5. The count register is not loaded until the count value is written (one or two bytes, depending on the mode selected), followed by a rising edge and a falling edge of the clock. Any read of the counter prior to that falling edge may yield invalid data.

Write Operations

The programming procedure for the 8254 is very flexible. There are but two conventions to be remembered:

1. For each counter, the control word must be written before the initial count is written.

2. The initial count must follow the count format specified in the control word, which is least significant byte (LSB) only, most significant byte (MSB) only, or least significant followed by most significant.

Since the control word and each of the three counters has its own addressing—selected by the A1, A0 inputs and each control word specificies the counter it applies to (SC0,SC1 bits)—no special instruction sequence is required. Any programming sequence that follows the preceding conventions is acceptable. *Figure 10.6* provides four possible programming sequences.

A new initial count may be written to a counter at any time without affecting the counter's programmed mode in any respect. The new count must, of course, follow the existing programmed count format.

If a counter is programmed to read/write two-byte counts, the following precaution applies: A program must not transfer control between writing the first and second byte to another routine, which also writes into the same counter. In that event, the counter will be loaded with an incorrect count.

Figure 10.7. 8254/8254-2 Counter Latching Command programming format. No requirement for a specific programming sequence exists. One, two or all three counters may be programmed for operation as required.

```
A1,A0=0;  CS=0;  RD=1;  WR=0
```

D7	D6	D5	D4	D3	D2	D1	D0
SC1	SC0	0	0	x	x	x	x

SC1,SC0 — specify counter to be latched.

SC1	SC0	COUNTER
0	0	0
0	1	1
1	0	2
1	1	Read Back Command

D5,D4 — 00 designates counter latching operation.

x — Don't care

Don't care bits should be 0 to ensure compatibility with future Intel products.

Read Operations

It is easy to read the value of a counter in the 8254 without disturbing the count in progress. There are three possible methods for this. The first is through the read-back command. The second is a simple read operation of the counter, which is selected by the A1,A0 inputs. The only requirement is that (1) the clock input of the selected counter must be inhibited, either by using the gate input or external logic, or (2) the count must first be latched. Otherwise, the count might be in the process of changing when it is read, leading to an undefined result.

Counter Latch Command

This method involves a special software command called the counter latch command. Like a control word, this command is written to the control word register, which is selected when A1,A0=11. Also, like a control word, the SC0, SC1 bits select one of the three counters, but the two other bits (D5 and D4) distinguish the command from a control word. The format for the counter-latch command is shown in *Figure 10.7*.

The selected counter's output latch (OL) latches the count at the time the counter latch command is received. This count is held in the latch until it is read by the CPU (or until the counter is reprogrammed). The count is then unlatched automatically and the OL returns to "following" the counting element (CE). This method allows reading the counters "on the fly"

Figure 10.8. 8254/8254-2 Read-Back Command format. The command applies to the Counters selected by setting their corresponding bits D3, D2, D1 = 1.

A1,A0=0; $\overline{CS}$=0; $\overline{RD}$=1; $\overline{WR}$=0							
D7	D6	D5	D4	D3	D2	D1	D0
1	1	COUNT	STATUS	CNT 2	CNT 1	CNT 0	0

```
D5: 0 = LATCH COUNT OF SELECTED COUNTER(S)
D4: 0 = LATCH STATUS OF SELECTED COUNTER(S)
D3: 1 = SELECT COUNTER 2
D2: 1 = SELECT COUNTER 1
D1: 1 = SELECT COUNTER 0
D0: RESERVED FOR FUTURE EXPANSION
```

D7	D6	D5	D4	D3	D2	D1	D0
OUTPUT	NULL COUNT	RW1	RW0	M2	M1	M0	BCD

```
D7 1 = OUT PIN IS 1
   0 = OUT PIN IS 0
D6 1 = NULL COUNT
   0 = COUNT AVAILABLE FOR READING
D5-D0 = COUNTER PROGRAMMED MODE
```

Figure 10.9. 8254/8254-2 Status Byte format. Bits D5 through D0 contain the Counter's programmed bits as written in the last Mode Control Word.

without affecting the count in progress. Multiple counter-latch commands may be used to latch more than a single counter. Each counter's OL holds count until it is read. Counter-latch commands don't affect the mode of the counter in any way.

If a counter is latched and then, some time later, latched again before the count is read, the second counter-latch command is ignored. The count read will be the count at the time the first counter-latch command was issued.

With either method, the count must be read according to the programmed format. Specifically, if the counter is programmed for two-byte counts, then two bytes must be read. The two bytes do not have to be read one right after the other. Read or write or programming operations of other counters may be inserted between them.

Another feature of the 8254 is that reads and writes of the same counter may be interleaved. For example, if the counter is programmed for two-byte counts, the following sequence is valid.

1. Read least significant byte.
2. Write new least significant byte.
3. Read most significant byte.
4. Write new most significant byte.

```
        THIS ACTION                                    CAUSES:
   A.   Write to the Control Word Register (1)         Null Count=1
   B.   Write to the Count Register (CR) (2)           Null Count=1
   C.   New Count is loaded into CE (CR→CE)            Null Count=0

  (1)   Only the counter specified by the Control Word will have
        its Null Count set to 1, Null Count Bits of other counters
        are unaffected.

  (2)   If the Counter is programmed for two-byte counts (Least
        Significant Byte the Most Significant Byte) Null Count goes
        to 1 when the second byte is written.
```

Figure 10.10. 8254/8254-2 Null Count Operation. Note that bit D6 of the Status Byte is the Null Count selector.

If a counter is programmed to read/write two-byte counts, the following pre-caution applies. A program must not transfer control between reading the first and second byte to another routine that also reads from the same counter. Otherwise, an incorrect count will be read.

Read-Back Command

The read-back command allows the user to check the count value, pro-grammed mode, and current state of the out pin and null-count flag of the selected counter(s).

The command is written into control word register and has the format shown in *Figure 10.8*. The command applies to the counters selected by setting their corresponding bits D3,D2,D1=1.

The read-back command may be used to latch multiple counter output latches (OL) by setting the /COUNT bit D5=0 and selecting the desired counters. This single command is functionally equivalent to several counter latch commands, one for each counter latched. Each counter's latched count is held until it is read (or the counter is reprogrammed). That counter is automatically unlatched when read, but other counters remain latched until they are read. If multiple read-back commands are issued to the same counter without reading the count, all but the first are

COMMAND								DESCRIPTION	RESULT
D7	D6	D5	D4	D3	D2	D1	D0		
1		0	0	0	0	1	0	Read back count and status of Counter 0	Count and status latched for Counter 1
1		1	0	0	1	0	0	Read back status of Counter 1	Status latched for Counter 1
1		1	0	1	1	0	0	Read back status of Counters 1,2	Status latched for Counter 2, but not Counter 1
1		0	1	1	0	0	0	Read back count of Counter 2	Count latched for Counter 2
1		0	0	0	1	0	0	Read back count and status of Counter 1	Count latched for Counter 1 but not status
1		1	0	0	0	1	0	Read back status of Counter 1	Command ignored, status already latched for Counter 1

Figure 10.11. 8254/8254-2 Read-Back Command example. Both Count and Status of the selected counter(s) may be latched simultaneously by setting both /COUNT and /STATUS bits D5,D4=0.

ignored—that is, the count that will be read is the count that existed at the time the first read-back command was received.

The read-back command may also be used to latch status information of selected counter(s) by setting /STATUS bit D4=0. Status must be latched to be read; status of a counter is accessed by a read from that counter.

The counter status format is shown in *Figure 10.9*. Bits D5 through D0 contain the counter's programmed mode exactly as written in the last mode control word. Output bit D7 contains the current state of the out pin. This allows the user to monitor the counter's output via software, possibly eliminating some hardware from a system.

Null-count bit D6 indicates when the last count written to the counter register (CR) has been loaded into the counting element (CE). The exact time this happens depends on the mode of the counter and is described in the mode definitions, but until the count is loaded into the CE it can't be read from the counter. If the count is latched or read before this time the count value will not reflect the new count just written. The operation of the null count is shown in *Figure 10.10*.

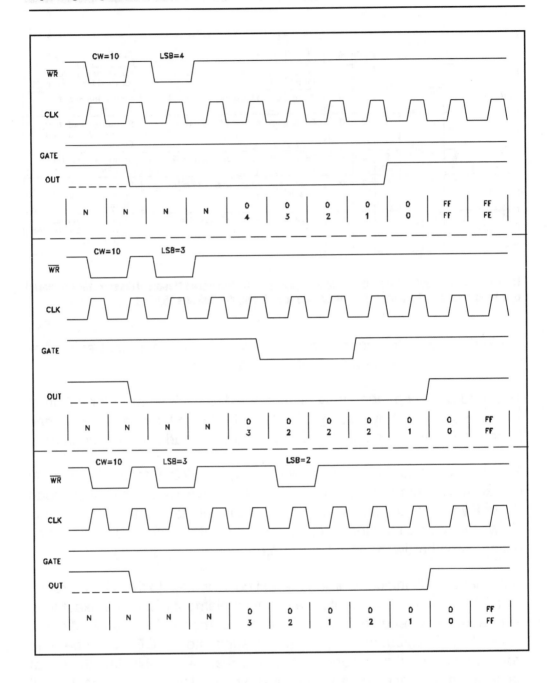

Figure 10.12. 8254/8254-2 Mode 0 Timing Diagram examples.

If multiple status latch operations of the counter(s) are performed without reading the status, all but the first are ignored—that is, the status that will be read is the status of the counter at the time the first status read-back was issued.

Both count and status of the selected counter(s) may be latched simultaneously by setting both the /COUNT and /STATUS bits D5,D4=0. This is functionally the same as issuing two separate read-back commands at once, and the preceding applies here also. Specifically, if multiple count and/or status read-back commands are issued to the same counter(s) without any intervening reads, all but the first are ignored. This is illustrated in *Figure 10.11*.

If both count and status of a counter are latched the first read operation will return latched status regardless of which was issued first. The next one or two reads (depending on whether the counter was programmed for one or two-byte counts) return latched count. Subsequent reads return unlatched count.

Mode Definitions

The following are defined for use in describing the operation of the 8254.

CLK pulse: A rising edge, then a falling edge—in that order—of a counters CLK input.

Trigger: A rising edge of a counter's gate input.

Counter loading: The transfer of a count from the CR to the CE.

The following conventions apply to the timing diagrams for the six operating modes, shown in Figures 10.12 through 10.17.

1. Counters are programmed for binary, not BCD, counting and for reading/writing the LSB only.

2. The counter is always selected, /CS low.

3. CW stands for "control word." CW=10 means a control word of 10. HEX is written to the counter.

4. LSB stands for "least significant byte" of count.

5. Numbers below diagrams are count values. The lower number is the LSB. The upper number is the MSB. Since the counter is programmed to read/write LSB only the MSB cannot be read. N stands for an undefined count. Vertical lines show transitions between count values.

Each of the mode diagrams illustrates three conditions.

Mode 0: Interrupt on Terminal Count

This mode is typically used for event counting. *Figure 10.12* is the timing diagram. The counter output will be remain low following the mode set operation. When the count has been loaded into the selected count register the output continues to remain low until the count has decremented to zero at which time the output goes high. The output will remain high until the selected count register is loaded with a new mode or a new count.

Gate=1 enables counting; gate=0 disables counting. Gate has no effect on OUT.

After the control word and initial count are written to a counter, the initial count will be loaded on the next CLK pulse. This CLK pulse does not decrement the count, so for an initial count of N, OUT does not go high until N+1 CLK pulses after the initial count is written.

If a new count is written to the counter it will be loaded on the next CLK pulse and counting will continue from the new count. If a two-byte count is written, the following occurs.

1. Writing the first byte disables counting. OUT is set low immediately (no clock pulse required).

2. Writing the second byte allows the new count to be loaded on the next CLK pulse.

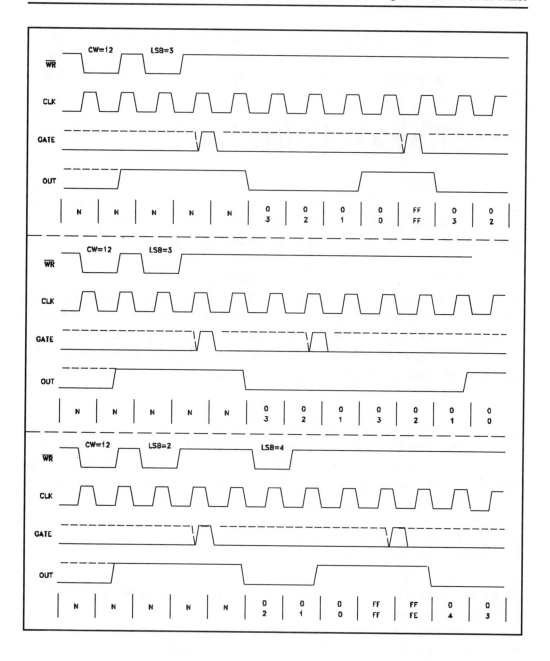

Figure 10.13. 8254/8254-2 Mode 1 Timing Diagram examples.

If an initial count is written while gate=0, it will still be loaded on the next CLK pulse. When gate goes high, OUT will go high N CLK pulses later, no CLK pulse is needed to load the counter as this has already been done.

Mode 1: Programmable One-Shot

OUT will be initially high. The counter output will go low on the count following the rising edge of the gate input trigger to begin the one-shot pulse. It will remain low until the counter reaches zero. OUT will then go high and remain high until the CLK pulse after the next trigger.

After writing the control word and initial count the counter is armed. A trigger results in loading the counter and setting OUT low on the next CLK pulse, thus starting the one-shot. An initial count of N will result in a one-shot pulse N CLK cycles in duration. The one-shot is retriggerable, hence OUT will remain low for N CLK cycles after any trigger. The pulse can be repeated without rewriting the same count into the counter. Gate has no effect on OUT.

If a new count is written into the counter during a one-shot pulse, the current one-shot is not affected unless the counter is retriggered. In that case the counter is loaded with the new count and the one-shot pulse continues until the new count expires. *Figure 10.13* illustrates the one-shot timing mode.

Mode 2: Rate Generator

This mode is that of a divide-by-N counter. It is typically used to generate a real-time clock interrupt. OUT will initially be high. When the initial count has decremented to 1, OUT goes low for one CLK pulse. OUT then goes high again, the counter reloads the initial count and the process repeats. Mode 2 is periodic; the same sequence repeats indefinitely. For an initial count of N the sequence repeats for every N CLK cycles.

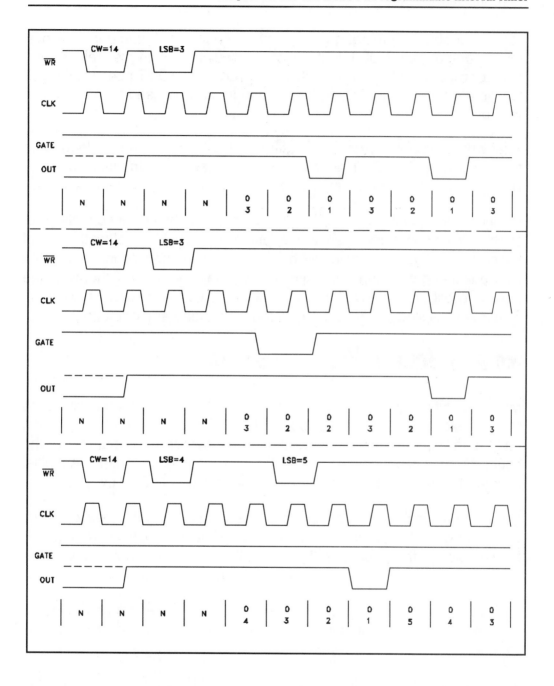

Figure 10.14. 8254/8254-2 Mode 2 Timing Diagram examples. Note that a Gate transition should not occur one clock cycle prior to terminal count.

Gate=1 enables counting; gate=0 disables counting. If gate goes low during an output pulse OUT is set high immediately. A trigger reloads the counter with the initial count on the next CLK pulse; OUT goes low N CLK pulses after the trigger. Thus, the gate input can be used to synchronize the counter.

After writing a control word and initial count, the counter will be loaded on the next CLK pulse. OUT goes low N CLK pulses after the initial count is written. This allows the counter to be synchronized by software also.

Writing a new count while counting does not affect the current counting sequence. If a trigger is received after writing a new count but before the end of the current period the counter will be loaded with the new count on the next CLK pulse and counting will continue from the new count. Otherwise, the new count will be loaded at the end of the current counting cycle. In Mode 2, a count of 1 is illegal. *Figure 10.14* is the timing diagram for Mode 2.

Mode 3: Square-Wave Generator

Mode 3 is typically used for baud-rate generation. This mode is similar to Mode 2 except for the duty cycle of OUT. OUT will initially be high. When half the initial count has expired, OUT goes low for the remainder of the count. Mode 3 is periodic; the sequence is repeated indefinitely. An initial count of N results in a square wave with a period of N CLK cycles.

Gate=1 enables counting; gate=0 disables counting. If gate goes low while OUT is low, OUT is set high immediately, and no CLK pulse is required. A trigger reloads the counter with the initial count on the next CLK pulse. Thus the gate input can be used to synchronize the counter.

After writing a control word and initial count, the counter will be loaded on the next CLK pulse. This allows the counter to be synchronized by software also.

Writing a new count while counting does not affect the current counting sequence. If a trigger is received after writing a new count but before the end of the current half-cycle of the square wave the counter will be loaded

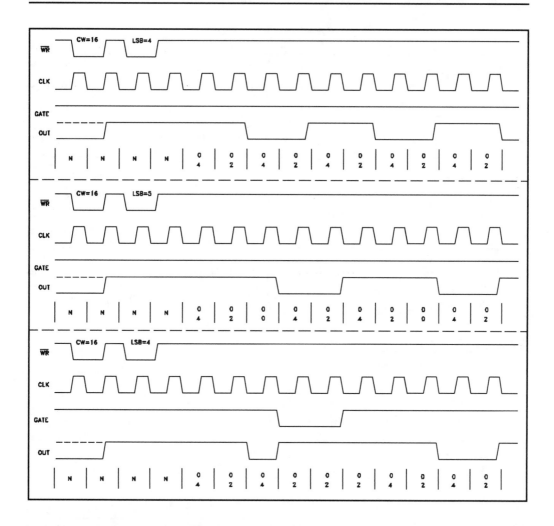

Figure 10.15. 8254/8254-2 Mode 3 Timing Diagram examples. Note that a Gate transition should not occur one clock cycle prior to terminal count.

with the new count on the next CLK pulse and counting will continue from the new count. Otherwise, the new count will be loaded at the end of the current half-cycle.

Mode 3 is implemented as follows.

Even counts: OUT is initially high. The initial count is loaded on one CLK pulse and then is de-cremented by two on succeeding CLK pulses. When

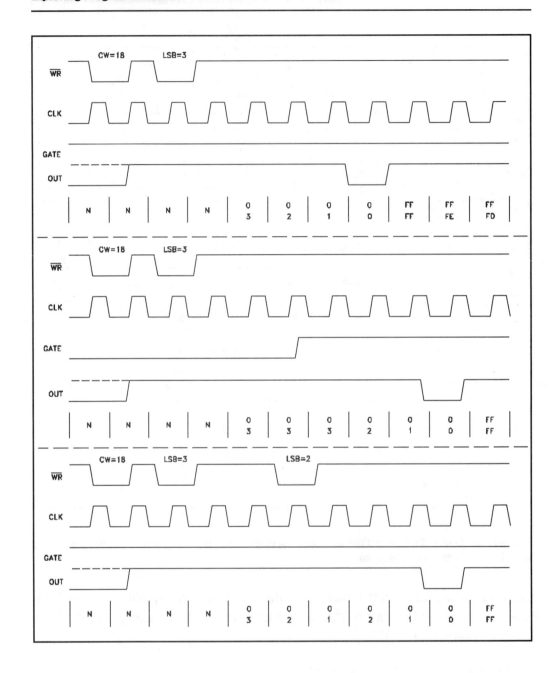

Figure 10.16. 8254/8254-2 Mode 4 Timing Diagram examples.

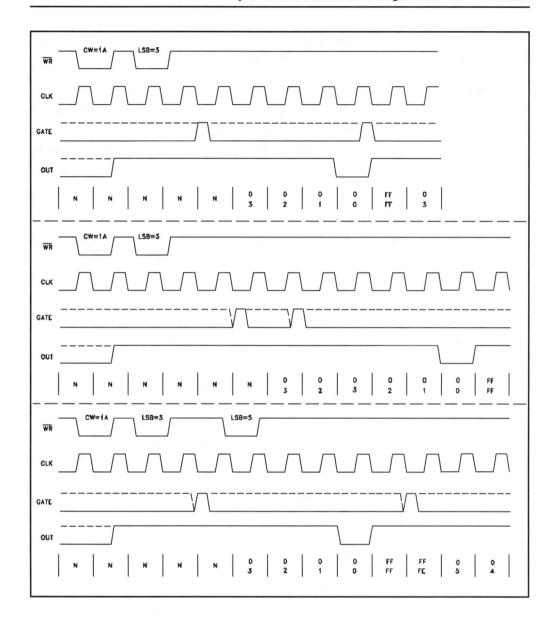

Figure 10.17. 8254/8254-2 Mode 5 Timing Diagram examples.

the count expires, OUT changes value and the counter is reloaded with the initial count. The process is repeated indefinitely.

Odd counts: OUT is initially high. The initial count minus one (an even number) is loaded on one CLK pulse and then is de-cremented by two on succeeding CLK pulses. One CLK pulse *after* the count expires, OUT goes low and the counter is reloaded with the initial count minus one. Succeeding CLK pulses decrement the count by two.

When the count expires, OUT goes goes high again and the counter is reloaded with the initial count minus one. The process is repeated indefinitely. So for odd counts OUT will be high for (N+1)/2 counts and low for (N-1)/2 counts. *Figure 10.15* is the timing diagram for Mode 3.

Modes	Signal Status	Low or Going Low	Rising	High
0		Disables Counting	---	Enables Counting
1		---	1) Initiates Counting 2) Resets after Next Clock	---
2		1) Disables Counting 2) Sets Output Immediately High	1) Reloads Counter 2) Initiates Counting	Enables Counting
3		1) Disables Counting 2) Sets Output Immediately High	1) Reloads Counter 2) Initiates Counting	Enables Counting
4		Disables Counting	---	
5		---	Initiates Counting	---

Figure 10.18. 8254-5/8254-2 gate functions for each of the six modes of counter operation. Note that both level and rise/fall transition requirements exist.

Mode	Min Count	Max Count
0	1	0
1	1	0
2	2	0
3	2	0
4	1	0
5	1	0

NOTE: 0 is equivalent to 10^{16} for Binary Counting and 10^4 for BCD.

Figure 10.19. Minimum and maximum initial counts for the 8254/8254-2.

Mode 4: Software-Triggered Strobe

OUT will be initially high. When the initial count expires, OUT will go low for one CLK pulse and then go high again. The counting sequence is "triggered" by writing the initial count.

Gate=1 enables counting; gate=0 disables counting. Gate has no effect on OUT.

After writing a control word and initial count, the counter will be loaded on the next CLK pulse. This CLK pulse does not decrement the count, so for an initial count of N OUT does not strobe low until N+1 CLK pulses after the initial count is written.

If a new count is written during counting, it will be loaded on the next CLK pulse and counting will continue from the new count. If a two-byte count is written, the following occurs.

1. Writing the first byte has no effect on the counting.

2. Writing the second byte allows the new count to be loaded on the next CLK pulse.

This allows the sequence to be "retriggered" by software. OUT strobes low N+1 CLK pulses after the new count of N is written. *Figure 10.16* is the timing diagram for Mode 4.

Bus Parameters 1

READ CYCLE

Symbol	Parameter	8254		8254-2		Unit
		Min.	Max.	Min.	Max.	
tAR	Address Stable Before RD↓	45		30		nS
tSR	CS Stable Before RD↓	0		0		nS
tRA	Address Hold Time After RD↓	0		0		nS
tRR	RD Pulse Width	150		95		nS
tRD	Data Delay from RD↓		120		85	nS
Tad	Data Delay from Address		220		185	nS
tDF	RD to Data Floating	5	90	5	65	nS
tRV	Command Recovery Time	200		165		nS

WRITE CYCLE

Symbol	Parameter	8254		8254-2		Unit
		Min.	Max.	Min.	Max.	
tAW	Address Stable Before WR↓	0		0		nS
tSW	CS Stable Before WR↓	0		0		nS
tWA	Address Hold Time WR↓	0		0		nS
tWW	WR Pulse Width	150		95		nS
tDW	Data Setup Time Before WR↓	120		95		nS
tWD	Data Hold Time After WR↓	0		0		nS
tRV	Command Recovery Time	200		165		nS

CLOCK AND GATE TIMING

Symbol	Parameter	8254		8254-2		Unit
		Min.	Max.	Min.	Max.	
tCLK	Clock Period	125	DC	100	DC	nS
tPHW	High Pulse Width	60 [3]		30 [3]		nS
tPWL	Low Pulse Width	60 [3]		50 [3]		nS
tR	Clock Rise Time		25		25	nS
tF	Clock Fall Time		25		25	nS
tGW	Gate Width High	50		50		nS
tGL	Gate Width Low	50		50		nS
tGS	Gate Set Up Time to Clock↑	50		40		nS
tGH	Gate Hold Time After Clock↑	50 [2]		50 [2]		nS
tOD	Output Delay From Clock↓		150		100	nS
tODG	Output Delay From Gate ↓		120		100	nS
tWC	CLK Delay for Loading	0	55	0	55	nS
tWG	Gate Delay for Sampling	−5	50	−5	40	nS
tWO	OUT Delay from Mode Write		260		240	nS
tCL	CLK Set Up for Count Latch	−40	45	−40	40	nS

NOTES:
1. AC timings measured at VOH=2.2, VOL=0.8
2. In modes 1 and 5 triggers are sampled on each rising clock edge. A second trigger with 120 nS (70 nS for 8254-2) of the rising clock edge may not be detected.
3. Low going glitches that violate tPWL may cause errors rquiring counter reprogramming.

Table 10.1. 8254/8254-2 AC characteristics.

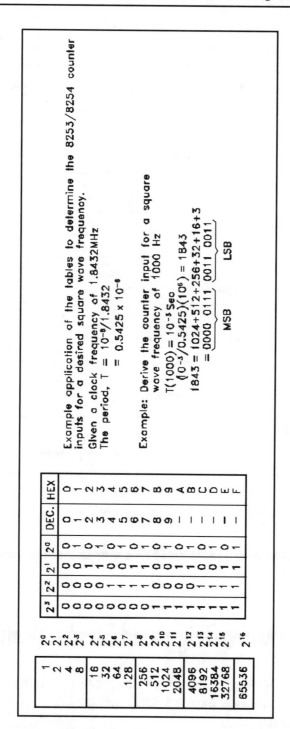

Example application of the tables to determine the 8253/8254 counter inputs for a desired square wave frequency.

Given a clock frequency of 1.8432MHz
The period, $T = 10^{-6}/1.8432$
$= 0.5425 \times 10^{-6}$

Example: Derive the counter input for a square wave frequency of 1000 Hz

$T(1000) = 10^{-3}$ Sec

$(10^{-3}/0.5425)(10^{6}) = 1843$

$1843 = 1024+512+256+32+16+3$

$= \underbrace{0000\ 0111}_{MSB}\ \underbrace{0011\ 0011}_{LSB}$

2^3	2^2	2^1	2^0	DEC.	HEX
0	0	0	0	0	0
0	0	0	1	1	1
0	0	1	0	2	2
0	0	1	1	3	3
0	1	0	0	4	4
0	1	0	1	5	5
0	1	1	0	6	6
0	1	1	1	7	7
1	0	0	0	8	8
1	0	0	1	9	9
1	0	1	0	—	A
1	0	1	1	—	B
1	1	0	0	—	C
1	1	0	1	—	D
1	1	1	0	—	E
1	1	1	1	—	F

2^0	1
2^1	2
2^2	4
2^3	8
2^4	16
2^5	32
2^6	64
2^7	128
2^8	256
2^9	512
2^{10}	1024
2^{11}	2048
2^{12}	4096
2^{13}	8192
2^{14}	16384
2^{15}	32768
2^{16}	65536

Table 10.2.   Binary and BCD tables with a worked example for a Counter input to provide a square-wave output.

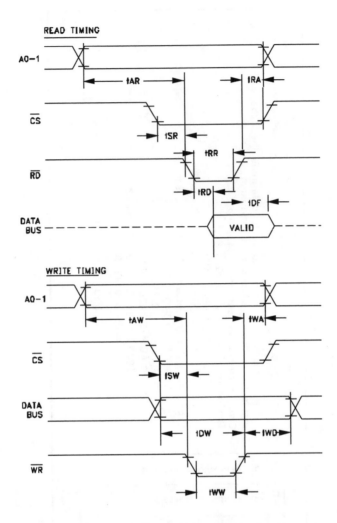

Figure 10.20. Timing diagrams for the 8254/8254-2 Read, Write, Clock and Gate, and Recovery operations.
(Continued on next page.)

Mode 5: Hardware-Triggered Strobe (Retriggerable)

OUT will initially be high. Counting is triggered by a rising edge of gate. When the initial count has expired, OUT will go low for one CLK pulse and then go high again.

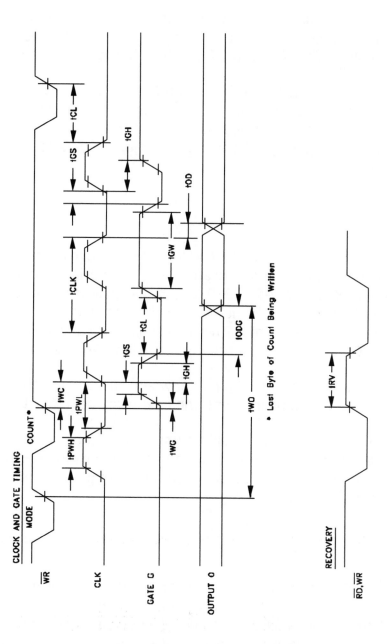

After writing the control word and initial count, the counter will not be loaded until the CLK pulse after a trigger. This CLK pulse does not decrement the count, so for an intial count of N OUT does not strobe low until N+1 CLK pulses after a trigger.

A trigger results in the counter being loaded with the initial count on the next CLK pulse. The counting sequence is retriggerable. OUT will not strobe low for N+1 CLK pulses after after any trigger. Gate has no effect on OUT.

If a new count is written during counting the current counting sequence will not be affected. If a trigger occurs after the new count is written but before the new count expires, the counter will be loaded with the new count on the next CLK pulse and counting will continue as before. *Figure 10.17* is a timing diagram for Mode 5.

Gate

The gate input is always sampled on the rising edge of CLK. In Modes 0, 2, 3, and 4, the gate input is level sensitive and the logic level is sampled on the rising edge of CLK. In Modes 1, 2, 3, and 5, the gate input is rising-edge sensitive. In these modes, a rising edge of gate (trigger) sets an edge-sensitive flip-flop in the counter. This flip-flop is then sampled on the next rising edge of CLK; the flip-flop is reset immediately after it is sampled.

In this way a trigger will be detected no matter when it occurs: a high logic level does not have to be maintained until the next rising edge of CLK.

Note that in Modes 2 and 3, if a CLK source other than the system clock is used, gate should be pulsed immediately following /WR of a new count value. *Figure 10.18* provides a summary of gate pin operations for each of the modes.

Counter

New counts are loaded and counters are decremented on the falling edge of CLK. The largest possible initial count is 0; this is the equivalent of 2e16 for binary counting and 10e4 for BCD counting. Minimum and maximum counts for each mode are shown in *Figure 10.19*.

The counter does not stop when it reaches zero. In Modes 0, 1, 4, and 5, the counter "wraps around" to the highest count—either FFFF for binary counting or 9999 for BCD counting—and continues with the count. Modes

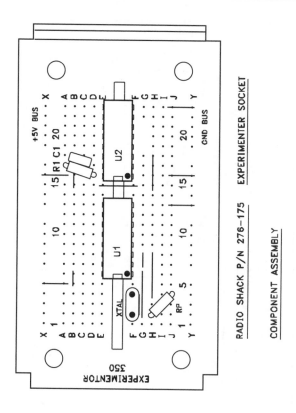

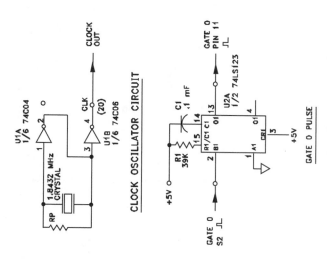

Figure 10.21.   The 1.8432-MHz clock source and Gate 0 pulse-source circuitry and prototype assembly used with the bench-top operation of the 8254/8254-2.

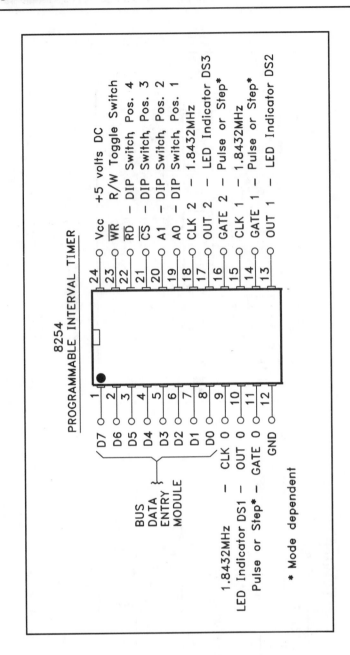

Figure 10.22. A helpful planning diagram for wiring the bench-top configuration of the 8254/8254-2.

2 and 3 are periodic; the counter reloads itself with the initial count and continues counting from there.

8254 Timing and AC Characteristics

Table 10.1 defines the AC characteristics for the 8254 and the 8254-2. *Figure 10.20* provides timing diagrams for read, write and clock, gate, and recovery timing.

Bench-Top Operation of the 8254/8254-2 Programmable Interval Timer

The 8254/8254-2 programmable interval timer is a departure from the data communication devices of the earlier chapters. As we can see from the bench-top assembly drawings it is a simpler and easier device to configure for operation. Operation is also very straightforward, as we will discover from the procedures provided.

There is, however, a complexity of another sort in the programming of the counters. In this we have the option of binary or BCD notation. *Table 10.2* provides a binary table plus an example of a square-wave frequency application.

Let's consider the basics now. The clock source is a frequency of 1.8432 MHz. The period of one clock cycle is:

$$10(e-6)/1.8432 = 0.5425(10e-6) = 0.5425 \ \mu S$$

Now suppose we want to use Counter 1 in the one-shot mode with a pulse width of 330 µS. The ratio of 330/.5425 = 608.3, rounded off to 608.

Let's take advantage of the binary table:

$$608 = 512+64+32$$
$$= 0000+0010+0110+0000$$

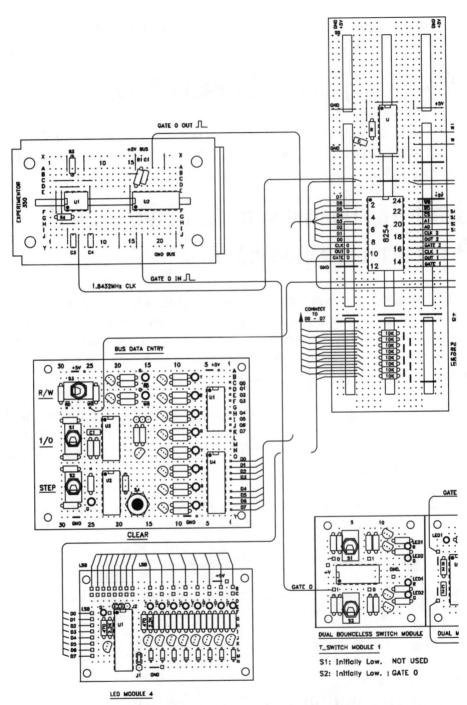

Figure 10.23. The author's assembly and wiring for bench-top operation of the 8254/8254-2. *(Continued on next page.)*

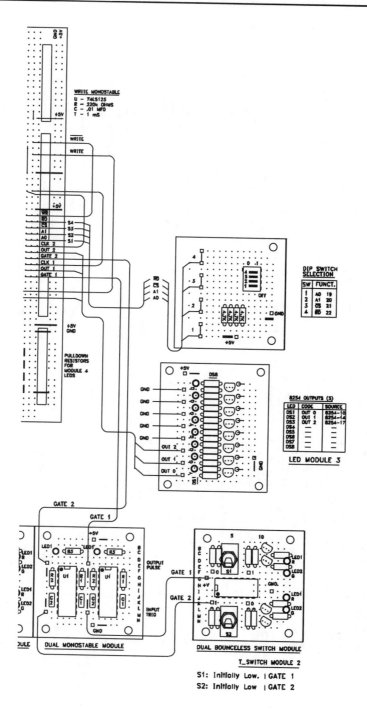

PROCEDURE FOR 8254 OPERATION AS A SQUARE WAVE GENERATOR

NOTES: The Bus Data Entry Toggle Switch is to remain in the READ Position except when toggled for a WRITE operation. DIP Switch #4 to remain in the ↑ position except when a Data Bus READ is desired.
For this MODE Gate 0 is to be tied to the +5V bus.
An oscilloscope and frequency meter are required to properly observe the square wave timing and frequency.

1. Setup Counter 0 for the Square Wave, MODE 3.

BUS DATA ENTRY = 0011 0110

7	6	5	4	3	2	1	0	OPERATION
0	0	1	1	0	1	1	0	Counter and MODE select

16-Bit Binary Counter
Mode 3 select
Load LSB first, then MSB
Counter 0 select

Toggle Data Entry Switch to Write

DIP SW SETTINGS

2. Load Counter 0 with the Least Significant Bits (LSB)

BUS DATA ENTRY = 1010 1110

7	6	5	4	3	2	1	0	OPERATION
1	0	1	0	1	1	1	0	Least Significant Byte (LSB)

Binary 8+4+2
Binary 128+32
Note: Binary values shown are suggested only.
Toggle Data Entry Switch to Write

DIP SW SETTINGS

3. Load Counter 0 with the Most Significant Bits (MSB)

BUS DATA ENTRY = 0000 1101

7	6	5	4	3	2	1	0	OPERATION
0	0	0	0	1	1	0	1	Most Significant Byte (MSB)

Binary 2048+1024+256
Binary 0+0+0
Note: Binary values shown are suggested only.
The frequency is 520 Hz for the counter values given.
Toggle Data Entry Switch to Write

DIP SW SETTINGS

4. Load the counters, LSB and MSB, with all ones to observe a flickering of LED DS1, displaying OUT 0. With these values f ~ 29Hz, T ~ 34mS. Repeat steps 2 and 3 with new values. Keep in mind the LSB and MSB must both be entered, even if the MSB is all zeroes.

5. Setup Counter 0 for Counter Latch Command

BUS DATA ENTRY = 0000 0000

7	6	5	4	3	2	1	0	OPERATION
0	0	0	0	0	0	0	0	Counter 0 Latch Command

Don't Care, should be 0
00 designates Counter Latch
00 designates Counter 0

Toggle Data Entry Switch to Write

DIP SW SETTINGS

6. Read Counter 0 count with two successive Reads Position RD to "0" to Read, then return to "1".

Enter values read

7	6	5	4	3	2	1	0	OPERATION
								Least Significant Byte (LSB)
								Most Significant Byte (MSB)

DIP SW SETTINGS

7. Setup Counter 0 for Read-Back Command

BUS DATA ENTRY = 1100 0010

7	6	5	4	3	2	1	0	OPERATION
1	1	0	0	0	0	1	0	Read-Back Counter 0

Always 0
Select Counter 0
Latch Status
Latch Count
Read-Back Command

Toggle Data Entry Switch to Write

DIP SW SETTINGS

8. Read Counter 0 Status and Count with 3 successive reads Position RD to "0" to Read, then return to "1".

Enter values read

7	6	5	4	3	2	1	0	OPERATION
								Counter 0 Status
								Least Significant Byte (LSB)
								Most Significant Byte (MSB)

DIP SW SETTINGS

Figure 10.24. The procedure for bench top operation of the 8254/8254-2 as a square-wave generator using counter 0.

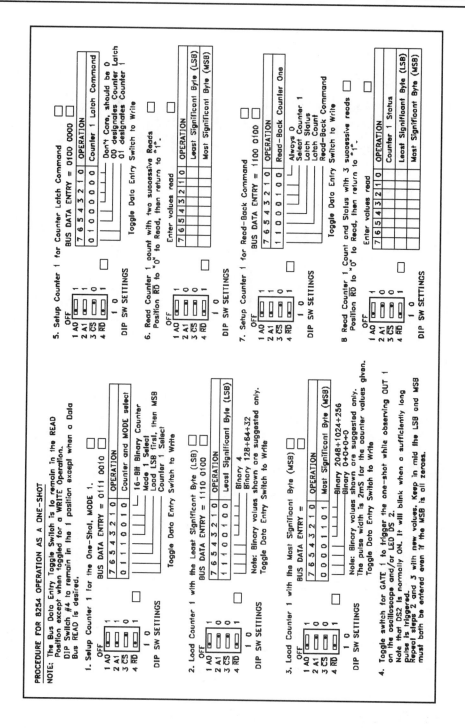

Figure 10.25. The procedure for bench-top operation of the 8254/8254-2 as a one-shot using counter 1.

Figure 10.26. **The procedure for bench-top operation of the 8254/8254-2 as a rate generator using Counter 2.**

in binary notation, from which the MSB is 0000 0010 and the LSB is 0110 0000.

We next enter these values, using the one-shot procedure of Figure 10.26. Upon triggering Gate 1, we see a negative-going pulse of 330 µS, just as we would expect. But first, we will have to learn how we get to this point in our experience.

Figure 10.21 describes the prototype clock circuitry and assembly employed in the previous chapters. So you may have made use of it already. There is a change in the monostable circuit in that only one is used, designated for the gate input of Counter 0. There is no change in the clock.

Figure 10.22 is a planning diagram for wiring the 8254 in the bench-top configuration. We will use it as a guide in wiring the operating setup, as shown in *Figure 10.23*.

Earlier chapters have provided suggestions for the construction of the modules and their corrugated cardboard supports so these are not repeated here. Do take care with the circuit connections, double-checking as you go, as with the congested layout it is easy to poke a wire into an incorrect point.

As with devices from the previous chapters, we are fortunate in that we are able to input data on the bus at a slow rate even though operation is with a high-speed clock. With the 8254, it is convenient to employ a monostable to clock in the data. We cannot take a similar route with reading the bus, however, as the data is not internally latched in these procedures. A DIP switch is used for /RD to enable reading a counter's status. Remember, though, that the control word register cannot be read.

Three test procedures are provided here, one for each counter. They are described in *Figures 10.24, 10.25,* and *10.26*.

The procedure of Figure 10.24 uses Counter 0 as a square-wave generator. For this function, we must connect the gate input to the +5V bus. The monostable connection shown in the wiring is in the event you wish to experiment with other applications of this counter.

Let's see where the binary values for the square wave came from.

1/520 = 1900 mS
1.900(10e-3)/.5425(10e-6) = 3502
From the binary table,
3502=2048 + 1024 + 256 + 128 + 32 + 8 + 4 + 2
    = 0000 + 1101 + 0011 + 0110
which is as shown in the procedure.

This procedure, and the two that follow, also includes steps for reading the counter status and count values with both the counter-latch and read-back commands. I have found the status values to be inconsistent, possibly because of the manner in which we are required to read the data. Count values will not repeat, as their value depends on the moment at which they are read.

The procedure of Figure 10.25 uses Counter 1 as a one-shot. In this we have worked out the math for a one-shot timing that differs from the example.

The procedure of Figure 10.26 uses Counter 2 as a rate generator. The counter output is high with the gate input low. If the gate is held high, continuous negative-going pulses will be outputted with a pulse width of the clock period. If the gate input is lowered, the pulses will cease, but raising the gate will restore them. Thus, the gate can be used as a synchronizer.

The procedures are designed to be self-explanatory as much as possible, but it will pay to review the text before proceeding with their operation. You may have difficulties, but perseverence pays dividends in satisfaction. After gaining experience with the examples given, you should experiment with other values for these modes and then venture forth with some of the other modes as well.

References

1. Intel Corporation, *Microsystem Components Handbook, Vol. 2,* Data Sheet, "8254," Programmable Interval Timer," 1984, p. 6-342.

2. Reference for Figure 10.1: Ibid. Figure 6, p. 6-345.

3. Reference for Figure 10.2:Ibid. Figures 1 and 2, p. 6-342.

4. Reference for Figure 10.3: Ibid. Figure 14, p. 6-349.

5. Reference for Figure 10.4: Ibid. Figure 5 p. 6-344.

6. Reference for Figure 10.5: Ibid. Figure 7, p. 6-345.

7. Reference for Figure 10.6: Ibid. Figure 8, p. 6-346.

8. Reference for Figure 10.7: Ibid. Figure 9, p. 6-347.

9. Reference for Figure 10.8: Ibid. Figure 10, p. 6-347.

10. Reference for Figure 10.9: Ibid. Figure 11, p. 6-348.

11. Reference for Figure 10.10: Ibid. Figure 12, p. 6-348.

12. Reference for Figure 10.11: Ibid. Figure 13, p. 6-348.

13. Reference for Figure 10.12: Ibid. Figure 15, p. 6-349.

14. Reference for Figure 10.13: Ibid. Figure 16, p. 6-350.

15. Reference for Figure 10.14: Ibid. Figure 17, p. 6-350.

16. Reference for Figure 10.15: Ibid. Figure 18, p. 6-351.

17. Reference for Figure 10.16: Ibid. Figure 19, p. 6-352.

18. Reference for Figure 10.17: Ibid. Figure 20, p. 6-352.

19. Reference for Figure 10.18: Ibid. Figure 21, p. 6-352.

20. Reference for Figure 10.19: Ibid. Figure 22, p. 6-353.

21. Reference for Figure 10.20: Ibid. Waveforms, p. 6-357.

22. Reference for Table 10.1: Ibid. AC Characteristics, p. 6-354, 6-355.

Index

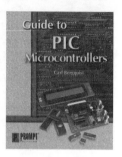

Exploring RF Circuits

Joseph J. Carr

Part the bushes cloaking the land of RF circuits and open the door to a world where you can learn all about RF circuitry testing, design, and construction. This book reveals practical RF circuitry projects that can be used and adapted to a variety of your needs and applications. All the projects have been user-tested by the author and other RF builders.

Detailed schematics, explanations, and tips will add ease and enjoyment to your journey, while you absorb the clarity, expertise, and years of electronics knowledge that author Joseph Carr offers as your guide.

Chapters cover varactor diodes, direct conversion receivers, RF signal generator circuits, RF grounding, RF bridges, fixed and variable capacitors, NE-602 chips, microwave-integrated circuits, radio transmission lines, and more. These projects that can be utilized by the hobbyist, student, or technician.

Modern Electronics
Soldering Techniques

Andrew Singmin

The traditional notion of soldering no longer applies in the quickly changing world of technology. Having the skills to solder electronics devices helps to advance your career.

Modern Electronics Soldering Techniques is designed as a total learning package, providing an extensive electronics foundation that enhances your electronics capabilities. This book covers how to solder wires and components as well as how to read schematics. Also learn how to apply your newly learned knowledge by following step-by-step instructions to take simple circuits and convert them into prototype breadboard designs. Other tospic covered include troubleshooting, basic math principles used in electronics, simple test meters and instruments, surface-mount technology, safety, and much more!

Electronics Technology
408 pages • paperback • 7-3/8" x 9-1/4"
ISBN: 0-7906-1197-X • Sams 61197
$34.95

Electronics Basics
304 pages • paperback • 6" x 9"
ISBN: 0-7906-1199-6 • Sams 61199
$24.95

To order today or locate your nearest Prompt® Publications distributor at 1-800-428-7267 or www.samswebsite.com

Prices subject to change.

Handbook for Parallel Port Design

James J. Barbarello

This book "demystifies" the parallel port. First addressing the basic tools of inputting to and outputting from the port, it logically progresses from the simple to more complex, showing how to use display devices (LEDs) and input sensing devices (such as light-sensitive resistors, IR LEDs, phototransistors, and rotary encoders). Then it walks you through the design process until you have an ample understanding of the parallel port and the skills to use it effectively.

A companion diskette is included, containing 52 support files, viewable in DOS, Windows® or Qbasic. These data and executable files will help you better understand and master the concepts presented, such as: Keycard Circuit Alignment; Rotary Encoder Demonstration; ADC Application; Computer Based Logic Probe Application; Executive Decision Maker Application; and others.

SMD Electronics Projects

Homer Davidson

Renowned author Homer Davidson has outdone himself by bringing you a book of 30 electronics projects, all utilizing surface-mounted devices. A fairly new technology, these projects are build with readily available components, and not only are they great fun to build, they can be of great use in your home.

Surface-mounted parts operate on low voltages and mount directly on the PC wiring. A parts list, schematic, wiring hookup, board layout, photos, and drawings help to illustrate each project. Troubleshooting procedures are given at the end of each project.

Projects include an Earphone Radio, Xtal Receiver, Shortwave Receiver, IC Radio, Shortwave Converter, FM Radio, Active Antenna, RF Amplifier, Sideband Adapter, Audio Amp, Xtal Earphones, and many, many more.

Electronics Technology
344 pages • paperback • 7-3/8 x 9-1/4"
ISBN: 0-7906-1177-5 • Sams: 61177
$29.95

Projects
320 pages • paperback • 7-3/8" x 9-1/4"
ISBN 0-7906-1211-9 • Sams 61211
$29.95

To order today or locate your nearest Prompt® Publications distributor at 1-800-428-7267 or www.samswebsite.com

Prices subject to change.